Shadowings

SHADOWINGS

by LAFCADIO HEARN

影

CHARLES E. TUTTLE COMPANY
Rutland, Vermont & Tokyo, Japan

Representatives

Continental Europe: BOXERBOOKS, INC., *Zurich*

British Isles: PRENTICE-HALL INTERNATIONAL, INC., *London*

Australasia: BOOK WISE (AUSTRALIA) PTY. LTD.
104-108 Sussex Street, Sydney 2000

Published by the Charles E. Tuttle Company, Inc.
of Rutland, Vermont & Tokyo, Japan
with editorial offices at
Suido 1-chome, 2-6, Bunkyo-ku, Tokyo, Japan

Library of Congress Catalog Card No. 77-138070

International Standard Book No. 0-8048-0967-4

First edition published 1900 by Little, Brown, and Co., Boston
First Tuttle edition published 1971
Third printing, 1980

0293-000259-4615
PRINTED IN JAPAN

Table of Contents

List of Illustrations

Publisher's Foreword

LAFCADIO HEARN is almost as Japanese as haiku. Both are an art form, an institution in Japan. Haiku is indigenous to the nation; Hearn became a Japanese citizen and married a Japanese, taking the name Yakumo Koizumi. His flight from Western materialism brought him to Japan in 1890. His search for beauty and tranquility, for pleasing customs and lasting values, kept him there the rest of his life, a Japanophile. He became the great interpreter of things Japanese to the West. His keen intellect, poetic imagination, and wonderfully clear style permitted him to penetrate to the very essence of things Japanese. In *Shadowings* Hearn delves deeply into Japanese life and comes up with some masterpieces of fact and fancy: warm, understandable glimpses into Japanese ways. Some of these bizarre and exotic tales are taken from such old Japanese story books as *Jikkun-sho* and *Konjaku Monogatari*.

Stories from Strange Books

Il avait vu brûler d'étranges pierres,
Jadis, dans les brasiers de la pensée . . .

Émile Verhaeren

The Reconciliation[1]

[1] The original story is to be found in the curious volume entitled *Konséki-Monogatari*

The Reconciliation

THERE was a young Samurai of Kyōto who had been reduced to poverty by the ruin of his lord, and found himself obliged to leave his home, and to take service with the Governor of a distant province. Before quitting the capital, this Samurai divorced his wife,—a good and beautiful woman,—under the belief that he could better obtain promotion by another alliance. He then married the daughter of a family of some distinction, and took her with him to the district whither he had been called.

But it was in the time of the thoughtlessness of youth, and the sharp experience of want, that the Samurai could not understand the worth of the affection so lightly cast away. His second marriage did not prove a happy one; the character of his new wife was hard and selfish; and he soon found every cause to think with regret of Kyōto days. Then he discovered that he still loved his first wife — loved her more than he could ever love the second; and he began to feel how unjust and how thankless he had been. Gradually his repentance deepened into a remorse that left him no peace of mind. Memories of the woman he had wronged — her gentle speech, her smiles, her dainty, pretty ways, her faultless patience — continually haunted him. Sometimes in dreams he saw her at her loom, weaving as when she toiled night and day to help him during the years of their distress: more often he saw her kneeling alone in the desolate little room where he had left her, veiling her tears with her poor worn sleeve. Even in the hours of official duty, his thoughts would wander back to her: then he would ask himself how she was living, what she was doing. Something in his heart assured him that she could not accept another husband, and that she never would refuse to pardon him. And he secretly resolved to seek her out as soon as he could return to Kyōto, — then to beg her forgiveness, to take her back, to do everything that a man could do to make atonement. But the years went by.

At last the Governor's official term expired, and the Samurai was free. "Now I will go back to my dear one," he vowed to himself. "Ah, what a cruelty,—what a folly to have divorced her!" He sent his second wife to her own people (she had given him no children); and hurrying to Kyōto, he went at once to seek his former companion,—not allowing himself even the time to change his travelling-garb.

When he reached the street where she used to live, it was late in the night,—the night of the tenth day of the ninth month;—and the city was silent as a cemetery. But a bright moon made everything visible; and he found the house without difficulty. It had a deserted look: tall weeds were growing on the roof. He knocked at the sliding-doors, and no one answered. Then, finding that the doors had not been fastened from within, he pushed them open, and entered. The front room was matless and empty: a chilly wind was blowing through crevices in the planking; and the moon shone through a ragged break in the wall of the alcove. Other rooms presented a like forlorn condition. The house, to all seeming, was unoccupied. Nevertheless, the Samurai determined to visit one other apartment at the further end of the dwelling,—a very small room that had been his wife's favorite resting-place. Approaching the sliding-screen that closed it, he was startled to perceive a glow within. He pushed the screen aside, and uttered a cry of joy; for he saw her there,—sewing by the light of a paper-lamp. Her eyes at the same instant met his own; and with a happy smile she greeted him,—asking only:—"When did you come back to Kyōto? How did you find your way here to me, through all those black rooms?" The years had not changed her. Still she seemed as fair and young as in his fondest memory of her;—but sweeter than any memory there came to him the music of her voice, with its trembling of pleased wonder.

Then joyfully he took his place beside her, and told her all:—how deeply he repented his selfishness,—how wretched he had been without her,—how constantly he had regretted her,—how long he had hoped and planned to make amends;—caressing her the while, and asking her forgiveness over and over again. She answered him, with loving gentleness, according to his heart's desire,—entreating him to cease all self-reproach. It was wrong, she said, that he should have allowed himself to suffer on her account: she had always felt that she was not worthy to be his wife. She knew that he had separated from her, notwithstanding, only because of poverty; and while he lived with her, he had always been kind; and she had never ceased to pray for his happiness. But even if there had been a reason for speaking of amends, this honorable visit would be ample amends;—what greater happiness than thus to see him again, though it were only for a moment? "Only for a moment!" he answered, with a glad laugh,—"say, rather, for the time of seven existences! My loved one, unless you forbid, I am coming back to live with you always—always—always! Nothing shall ever separate us again. Now I have means and friends: we need not fear poverty. To-morrow my goods will be brought here; and my servants will come to wait upon you; and we shall make this house beautiful. . . . To-night," he added, apologetically, "I came thus late—without even changing my dress—only because of the longing I had to see you, and to tell you this." She seemed greatly pleased by these words; and in her turn she told him about all that had happened in Kyōto since the time of his departure, — excepting her own sorrows, of which she sweetly refused to speak. They chatted far into the night: then she conducted him to a warmer room, facing south, — a room that had been their bridal chamber in former time. "Have you no one in the house to help you?" he asked, as she began to prepare the couch for him. "No," she answered, laughing cheerfully: "I could not afford a servant; — so I have been living all alone." "You will have plenty of servants to-morrow," he said, — "good servants, — and everything else that you need." They lay down to rest, — not to sleep: they had too much to tell each other; — and they talked of the past and the present and the future, until the dawn was grey. Then, involuntarily, the Samurai closed his eyes, and slept.

When he awoke, the daylight was streaming through the chinks of the sliding-shutters; and he found himself, to his utter amazement, lying upon the naked boards of a mouldering floor. . . . Had he only dreamed a dream? No: she was there; — she slept. . . . He bent above her, — and looked, — and shrieked; — for the sleeper had no face! . . Before him, wrapped in its grave-sheet only, lay the corpse of a woman, — a corpse so wasted that little remained save the bones, and the long black tangled hair.

. . . .

Slowly, — as he stood shuddering and sickening in the sun, — the icy horror yielded to a despair so intolerable, a pain so atrocious, that he clutched at the mocking shadow of a doubt. Feigning ignorance of the neighborhood, he ventured to ask his way to the house in which his wife had lived.

"There is no one in that house," said the person questioned. "It used to belong to the wife of a Samurai who left the city several years ago. He divorced her in order to marry another woman before he went away; and she fretted a great deal, and so became sick. She had no relatives in Kyōto, and nobody to care for her; and she died in the autumn of the same year, — on the tenth day of the ninth month. . ."

A Legend of Fugen-Bosatsu[1]

[1] From the old story-book, *Jikkun-shō*

A Legend of Fugen-Bosatsu

THERE was once a very pious and learned priest, called Shōku Shōnin, who lived in the province of Harima. For many years he meditated daily upon the chapter of Fugen-Bosatsu [the Bodhisattva Samantabhadra] in the Sûtra of the Lotos of the Good Law; and he used to pray, every morning and evening, that he might at some time be permitted to behold Fugen-Bosatsu as a living presence, and in the form described in the holy text.[1]

---

[1] The priest's desire was probably inspired by the promises recorded in the chapter entitled "The Encouragement of Samantabhadra" (see Kern's translation of the Saddharma Pundarîka in the *Sacred Books of the East,*—pp. 433–434):—"Then the Bodhisattva Mahâsattva Samantabhadra said to the Lord: . . . 'When a preacher who applies himself to this Dharmaparyâya shall take a walk, then, O Lord, will I mount a white elephant with six tusks, and betake myself to the place where that preacher is walking, in order to protect this Dharmaparyâya. And

One evening, while he was reciting the Sûtra, drowsiness overcame him; and he fell asleep leaning upon his *kyōsoku.*[1] Then he dreamed; and in his dream a voice told him that, in order to see Fugen-Bosatsu, he must go to the house of a certain courtesan, known as the "Yujō-no-Chōja,"[2] who lived in the town of Kanzaki. Immediately upon awakening he resolved to go to Kanzaki;—and, making all possible haste, he reached the town by the evening of the next day.

When he entered the house of the *yujō*, he found many persons already there assembled—mostly young men of the capital, who had been attracted to Kanzaki by the fame of the woman's

---

when that preacher, applying himself to this Dharma-paryâya, forgets, be it but a single word or syllable, then will I mount the white elephant with six tusks, and show my face to that preacher, and repeat this entire Dharma-paryâya."—But these promises refer to "the end of time."

[1] The *Kyōsoku* is a kind of padded arm-rest, or arm-stool, upon which the priest leans one arm while reading. The use of such an arm-rest is not confined, however, to the Buddhist clergy.

[2] A *yujō*, in old days, was a singing-girl as well as a courtesan. The term "Yujō-no-Chōja," in this case, would mean simply "the first (or best) of yujō."

beauty. They were feasting and drinking; and the *yujō* was playing a small hand-drum (*tsuzumi*), which she used very skilfully, and singing a song. The song which she sang was an old Japanese song about a famous shrine in the town of Murozumi; and the words were these:—

Within the sacred water-tank[1] *of Murozumi in*
Suwō,
Even though no wind be blowing,
The surface of the water is always rippling.

The sweetness of the voice filled everybody with surprise and delight. As the priest, who had taken a place apart, listened and wondered, the girl suddenly fixed her eyes upon him; and in the same instant he saw her form change into the form of Fugen-Bosatsu, emitting from her brow a beam of light that seemed to pierce beyond the limits of the universe, and riding a snow-white elephant with six tusks. And still

---

[1] *Mitarai. Mitarai* (or *mitarashi*) is the name especially given to the water-tanks, or water-fonts—of stone or bronze—placed before Shintō shrines in order that the worshipper may purify his lips and hands before making prayer. Buddhist tanks are not so named.

she sang — but the song also was now transformed; and the words came thus to the ears of the priest: —

On the Vast Sea of Cessation,
Though the Winds of the Six Desires and of the Five Corruptions never blow,
Yet the surface of that deep is always covered
With the billowings of Attainment to the Reality-in-Itself.

Dazzled by the divine ray, the priest closed his eyes: but through their lids he still distinctly saw the vision. When he opened them again, it was gone: he saw only the girl with her hand-drum, and heard only the song about the water of Murozumi. But he found that as often as he shut his eyes he could see Fugen-Bosatsu on the six-tusked elephant, and could hear the mystic Song of the Sea of Cessation. The other persons present saw only the *yujō*: they had not beheld the manifestation.

Then the singer suddenly disappeared from the banquet-room — none could say when or how. From that moment the revelry ceased; and gloom took the place of joy. After having waited and sought for the girl to no purpose, the company dispersed in great sorrow. Last of all, the priest departed, bewildered by the emotions of the evening. But scarcely had he passed beyond the gate, when the *yujō* appeared before him, and said:—"Friend, do not speak yet to any one of what you have seen this night." And with these words she vanished away,—leaving the air filled with a delicious fragrance.

***

The monk by whom the foregoing legend was recorded, comments upon it thus:—The condition of a *yujō* is low and miserable, since she is condemned to serve the lusts of men. Who therefore could imagine that such a woman might be the *nirmaṇakaya*, or incarnation, of a Bodhisattva. But we must remember that the Buddhas and the Bodhisattvas may appear in this world in countless different forms; choosing, for the purpose of their divine compassion, even the most humble or contemptible shapes when such shapes can serve them to lead men into the true path, and to save them from the perils of illusion.

The Screen-Maiden[1]

[1] Related in the *Otogi-Hyaku-Monogatari*

The Screen-Maiden

SAYS the old Japanese author, Hakubai-En Rosui: —[1]

"In Chinese and in Japanese books there are related many stories, — both of ancient and of modern times, — about pictures that were so beautiful as to exercise a magical influence upon the beholder. And concerning such beautiful pictures, — whether pictures of flowers or of birds or of people, painted by famous artists, — it is further told that the shapes of the creatures or

---

[1] He died in the eighteenth year of Kyōhō (1733). The painter to whom he refers — better known to collectors as Hishigawa Kichibei Moronobu — flourished during the latter part of the seventeenth century. Beginning his career as a dyer's apprentice, he won his reputation as an artist about 1680, when he may be said to have founded the *Ukiyo-yé* school of illustration. Hishigawa was especially a delineator of what are called *fūryū*, (" elegant manners "), — the aspects of life among the upper classes of society.

the persons, therein depicted, would separate themselves from the paper or the silk upon which they had been painted, and would perform various acts;—so that they became, by their own will, really alive. We shall not now repeat any of the stories of this class which have been known to everybody from ancient times. But even in modern times the fame of the pictures painted by Hishigawa Kichibei—'Hishigawa's Portraits'—has become widespread in the land."

He then proceeds to relate the following story about one of the so-called portraits:—

There was a young scholar of Kyōto whose name was Tokkei. He used to live in the street called Muromachi. One evening, while on his way home after a visit, his attention was attracted by an old single-leaf screen (*tsuitaté*), exposed for sale before the shop of a dealer in second-hand goods. It was only a paper-covered screen; but there was painted upon it the full-length figure of a girl which caught the young man's fancy. The price asked was very small: Tokkei bought the screen, and took it home with him.

When he looked again at the screen, in the solitude of his own room, the picture seemed to him much more beautiful than before. Apparently it was a real likeness, — the portrait of a girl fifteen or sixteen years old; and every little detail in the painting of the hair, eyes, eyelashes, mouth, had been executed with a delicacy and a truth beyond praise. The *manajiri*[1] seemed "like a lotos-blossom courting favor"; the lips were "like the smile of a red flower"; the whole young face was inexpressibly sweet. If the real girl so portrayed had been equally lovely, no man could have looked upon her without losing his heart. And Tokkei believed that she must have been thus lovely; — for the figure seemed alive, — ready to reply to anybody who might speak to it.

Gradually, as he continued to gaze at the picture, he felt himself bewitched by the charm of it. "Can there really have been in this world," he murmured to himself, "so delicious a creature? How gladly would I give my life — nay, a thousand years of life! — to hold her in my arms

---

[1] Also written *méjiri*, — the exterior canthus of the eye. The Japanese (like the old Greek and the old Arabian poets) have many curious dainty words and similes to express particular beauties of the hair, eyes, eyelids, lips, fingers. etc.

even for a moment!" (The Japanese author says "for a few seconds.") In short, he became enamoured of the picture, — so much enamoured of it as to feel that he never could love any woman except the person whom it represented. Yet that person, if still alive, could no longer resemble the painting: perhaps she had been buried long before he was born!

Day by day, nevertheless, this hopeless passion grew upon him. He could not eat; he could not sleep: neither could he occupy his mind with those studies which had formerly delighted him. He would sit for hours before the picture, talking to it, — neglecting or forgetting everything else. And at last he fell sick — so sick that he believed himself going to die.

Now among the friends of Tokkei there was one venerable scholar who knew many strange things about old pictures and about young hearts. This aged scholar, hearing of Tokkei's illness, came to visit him, and saw the screen, and understood what had happened. Then Tokkei, being questioned, confessed everything to his friend, and declared: — "If I cannot find such a woman, I shall die."

The old man said: —

"That picture was painted by Hishigawa Kichibei, — painted from life. The person whom it represented is not now in the world. But it is said that Hishigawa Kichibei painted her mind as well as her form, and that her spirit lives in the picture. So I think that you can win her."

Tokkei half rose from his bed, and stared eagerly at the speaker.

"You must give her a name," the old man continued; — "and you must sit before her picture every day, and keep your thoughts constantly fixed upon her, and call her gently by the name which you have given her, *until she answers you. . . .*"

"Answers me!" exclaimed the lover, in breathless amazement.

"Oh, yes," the adviser responded, "she will certainly answer you. But you must be ready, when she answers you, to present her with what I am going to tell you. . . ."

"I will give her my life!" cried Tokkei.

"No," said the old man; — "you will present her with a cup of wine that has been bought at one hundred different wine-shops. Then she will come out of the screen to accept the wine. After that, probably she herself will tell you what to do."

With these words the old man went away. His advice aroused Tokkei from despair. At once he seated himself before the picture, and called it by the name of a girl — (what name the Japanese narrator has forgotten to tell us) — over and over again, very tenderly. That day it made no answer, nor the next day, nor the next. But Tokkei did not lose faith or patience; and after many days it suddenly one evening answered to its name, —

"*Hai!*" (Yes.)

Then quickly, quickly, some of the wine from a hundred different wine-shops was poured out, and reverentially presented in a little cup. And the girl stepped from the screen, and walked upon the matting of the room, and knelt to take the cup from Tokkei's hand, — asking, with a delicious smile: —

"How could you love me so much?"

Says the Japanese narrator: "She was much more beautiful than the picture, — beautiful to the tips of her finger-nails, — beautiful also in heart and temper, — lovelier than anybody else in the world." What answer Tokkei made to her question is not recorded: it will have to be imagined.

"But will you not soon get tired of me?" she asked.

"Never while I live!" he protested.

"And after—?" she persisted;—for the Japanese bride is not satisfied with love for one life-time only.

"Let us pledge ourselves to each other," he entreated, "for the time of seven existences."

"If you are ever unkind to me," she said, "I will go back to the screen."

They pledged each other. I suppose that Tokkei was a good boy,—for his bride never returned to the screen. The space that she had occupied upon it remained a blank.

Exclaims the Japanese author,—

"How very seldom do such things happen in this world!"

The Corpse-Rider[1]

[1] From the *Konséki-Monogatari*

The Corpse-Rider

THE body was cold as ice; the heart had long ceased to beat: yet there were no other signs of death. Nobody even spoke of burying the woman. She had died of grief and anger at having been divorced. It would have been useless to bury her, — because the last undying wish of a dying person for vengeance can burst asunder any tomb and rift the heaviest graveyard stone. People who lived near the house in which she was lying fled from their homes. They knew that she was only *waiting for the return of the man who had divorced her*.

At the time of her death he was on a journey. When he came back and was told what had happened, terror seized him. "If I can find no help before dark," he thought to himself, "she will tear me to pieces." It was yet only the Hour of the Dragon;[1] but he knew that he had no time to lose.

He went at once to an *inyōshi*,[2] and begged for succor. The *inyōshi* knew the story of the dead woman; and he had seen the body. He said to the supplicant:—"A very great danger threatens you. I will try to save you. But you must promise to do whatever I shall tell you to do. There is only one way by which you can be saved. It is a fearful way. But unless you find the courage to attempt it, she will tear you limb from limb. If you can be brave, come to me again in the evening before sunset." The man shuddered; but he promised to do whatever should be required of him.

At sunset the *inyōshi* went with him to the house where the body was lying. The *inyōshi* pushed open the sliding-doors, and told his client to enter. It was rapidly growing dark. "I dare

---

1 *Tatsu no Koku*, or the Hour of the Dragon, by old Japanese time, began at about eight o'clock in the morning.

2 *Inyōshi*, a professor or master of the science of *in-yō*,—the old Chinese nature-philosophy, based upon the theory of a male and a female principle pervading the universe.

not!" gasped the man, quaking from head to foot;—"I dare not even look at her!" "You will have to do much more than look at her," declared the *inyōshi;*—"and you promised to obey. Go in!" He forced the trembler into the house and led him to the side of the corpse.

The dead woman was lying on her face. "Now you must get astride upon her," said the *inyōshi*, "and sit firmly on her back, as if you were riding a horse. . . . Come!—you must do it!" The man shivered so that the *inyōshi* had to support him—shivered horribly; but he obeyed. "Now take her hair in your hands," commanded the *inyōshi*,—"half in the right hand, half in the left. . . . So! . . . You must grip it like a bridle. Twist your hands in it—both hands—tightly. That is the way! . . . Listen to me! You must stay like that till morning. You will have reason to be afraid in the night—plenty of reason. But whatever may happen, never let go of her hair. If you let go,—even for one second,—she will tear you into gobbets!"

The *inyōshi* then whispered some mysterious words into the ear of the body, and said to its rider: — "Now, for my own sake, I must leave you alone with her. . . . Remain as you are! . . . Above all things, remember that you must not let go of her hair." And he went away, — closing the doors behind him.

Hour after hour the man sat upon the corpse in black fear; — and the hush of the night deepened and deepened about him till he screamed to break it. Instantly the body sprang beneath him, as to cast him off; and the dead woman cried out loudly, "Oh, how heavy it is! Yet I shall bring that fellow here now!"

Then tall she rose, and leaped to the doors, and flung them open, and rushed into the night, — always bearing the weight of the man. But he, shutting his eyes, kept his hands twisted in her long hair, — tightly, tightly, — though fearing with such a fear that he could not even moan. How far she went, he never knew. He saw nothing: he heard only the sound of her naked feet in the dark, — *picha-picha, picha-picha*, — and the hiss of her breathing as she ran.

At last she turned, and ran back into the house, and lay down upon the floor exactly as at first. Under the man she panted and moaned till the cocks began to crow. Thereafter she lay still.

But the man, with chattering teeth, sat upon her until the *inyōshi* came at sunrise. "So you did not let go of her hair!"—observed the *inyōshi*, greatly pleased. "That is well . . . Now you can stand up." He whispered again into the ear of the corpse, and then said to the man:—"You must have passed a fearful night; but nothing else could have saved you. Hereafter you may feel secure from her vengeance."

* * *

The conclusion of this story I do not think to be morally satisfying. It is not recorded that the corpse-rider became insane, or that his hair turned white: we are told only that "he worshipped the *inyōshi* with tears of gratitude." A note appended to the recital is equally disappointing. "It is reported," the Japanese author says, "that a grandchild of the man [*who rode the corpse*] still survives, and that a grandson of the *inyōshi* is at this very time living in a village called Otokunoi-mura [*probably pronounced Otonoi-mura*]."

This village-name does not appear in any Japanese directory of to-day. But the names of many towns and villages have been changed since the foregoing story was written.

The Sympathy of Benten[1]

[1] The original story is in the *Otogi-Hyaku-Monogatari*

The Sympathy of Benten

IN Kyōto there is a famous temple called Amadera. Sadazumi Shinnō, the fifth son of the Emperor Seiwa, passed the greater part of his life there as a priest; and the graves of many celebrated persons are to be seen in the temple-grounds.

But the present edifice is not the ancient Amadera. The original temple, after the lapse of ten centuries, fell into such decay that it had to be entirely rebuilt in the fourteenth year of Genroku (1701 A. D.).

A great festival was held to celebrate the rebuilding of the Amadera; and among the thousands of persons who attended that festival there was a young scholar and poet named Hanagaki Baishū. He wandered about the newly-laid-out grounds and gardens, delighted by all that he saw, until he reached the place of a spring at which he had often drunk in former times. He was then surprised to find that the soil about the spring had been dug away, so as to form a square pond, and that at one corner of this pond there had been set up a wooden tablet bearing the words *Tanjō-Sui* ("Birth-Water").[1] He also saw that a small, but very handsome temple of the Goddess Benten had been erected beside the pond. While he was looking at this new temple, a sudden gust of wind blew to his feet a *tanzaku*,[2] on which the following poem had been written:—

Shirushi arέto
Iwai zo somuru
Tama hōki,
Torutέ bakari no
Chigiri narέtomo.

This poem — a poem on first love (*hatsu koi*), composed by the famous Shunrei Kyō — was not

---

[1] The word *tanjō* (birth) should here be understood in its mystical Buddhist meaning of new life or rebirth, rather than in the western signification of birth.

[2] *Tanzaku* is the name given to the long strips or ribbons of paper, usually colored, upon which poems are written perpendicularly. Poems written upon *tanzaku* are suspended to trees in flower, to wind-bells, to any beautiful object in which the poet has found an inspiration.

unfamiliar to him; but it had been written upon the *tanzaku* by a female hand, and so exquisitely that he could scarcely believe his eyes. Something in the form of the characters,—an indefinite grace,—suggested that period of youth between childhood and womanhood; and the pure rich color of the ink seemed to bespeak the purity and goodness of the writer's heart.[1]

Baishū carefully folded up the *tanzaku*, and took it home with him. When he looked at it again the writing appeared to him even more wonderful than at first. His knowledge in caligraphy assured him only that the poem had been written by some girl who was very young, very intelligent, and probably very gentle-hearted.

---

[1] It is difficult for the inexperienced European eye to distinguish in Chinese or Japanese writing those characteristics implied by our term "hand"—in the sense of individual style. But the Japanese scholar never forgets the peculiarities of a handwriting once seen; and he can even guess at the approximate age of the writer. Chinese and Japanese authors claim that the color (quality) of the ink used tells something of the character of the writer. As every person grounds or prepares his or her own ink, the deeper and clearer black would at least indicate something of personal carefulness and of the sense of beauty.

But this assurance sufficed to shape within his mind the image of a very charming person; and he soon found himself in love with the unknown. Then his first resolve was to seek out the writer of the verses, and, if possible, make her his wife. . . . Yet how was he to find her? Who was she? Where did she live? Certainly he could hope to find her only through the favor of the Gods.

But presently it occurred to him that the Gods might be very willing to lend their aid. The *tanzaku* had come to him while he was standing in front of the temple of Benten-Sama; and it was to this divinity in particular that lovers were wont to pray for happy union. This reflection impelled him to beseech the Goddess for assistance. He went at once to the temple of Benten-of-the-Birth-Water (*Tanjō-sui-no-Benten*) in the grounds of the Amadera; and there, with all the fervor of his heart, he made his petition:—"O Goddess, pity me!—help me to find where the young person lives who wrote the *tanzaku!*—vouchsafe me but one chance to meet her,—even if only for a moment!" And after having made this prayer, he began to perform a seven days' religious service (*nanuka mairi*) [1] in honor of the Goddess; vowing at the same time to pass the seventh night in ceaseless worship before her shrine.

Now on the seventh night, — the night of his vigil, — during the hour when the silence is most deep, he heard at the main gateway of the temple-grounds a voice calling for admittance. Another voice from within answered; the gate was opened; and Baishū saw an old man of majestic appearance approaching with slow steps. This venerable person was clad in robes of ceremony; and he wore upon his snow-white head a black cap (*eboshi*) of the form indicating high rank. Reaching the little temple of Benten, he knelt down in front of it, as if respectfully awaiting some order. Then the outer door of the temple was opened; the hanging curtain of bamboo behind it, concealing the inner sanctuary, was rolled half-way up; and a *chigo* [2] came forward,

---

[1] There are many kinds of religious exercises called *mairi*. The performer of a *nanuka-mairi* pledges himself to pray at a certain temple every day for seven days in succession.

[2] The term *chigo* usually means the page of a noble household, especially an Imperial page. The *chigo* who

— a beautiful boy, with long hair tied back in the ancient manner. He stood at the threshold, and said to the old man in a clear loud voice: —

"There is a person here who has been praying for a love-union not suitable to his present condition, and otherwise difficult to bring about. But as the young man is worthy of Our pity, you have been called to see whether something can be done for him. If there should prove to be any relation between the parties from the period of a former birth, you will introduce them to each other."

On receiving this command, the old man bowed respectfully to the *chigo:* then, rising, he drew from the pocket of his long left sleeve a crimson cord. One end of this cord he passed round Baishū's body, as if to bind him with it. The other end he put into the flame of one of the temple-lamps; and while the cord was there burning, he waved his hand three times, as if to summon somebody out of the dark.

Immediately, in the direction of the Amadera, a sound of coming steps was heard; and in another

---

appears in this story is of course a supernatural being, — the court-messenger of the Goddess, and her mouthpiece.

moment a girl appeared, — a charming girl, fifteen or sixteen years old. She approached gracefully, but very shyly, — hiding the lower part of her face with a fan; and she knelt down beside Baishū. The *chigo* then said to Baishū: —

"Recently you have been suffering much heart-pain; and this desperate love of yours has even impaired your health. We could not allow you to remain in so unhappy a condition; and We therefore summoned the Old-Man-under-the-Moon[1] to make you acquainted with the writer of that *tanzaku*. She is now beside you."

With these words, the *chigo* retired behind the bamboo curtain. Then the old man went away as he had come; and the young girl followed him. Simultaneously Baishū heard the great bell of the Amadera sounding the hour of dawn. He prostrated himself in thanksgiving before the shrine of Benten-of-the-Birth-Water, and proceeded homeward, — feeling as if awakened from some delightful dream, — happy at having seen

---

[1] *Gekkawō.* This is a poetical appellation for the God of Marriage, more usually known as *Musubi-no-kami.* Throughout this story there is an interesting mingling of Shintō and Buddhist ideas.

the charming person whom he had so fervently prayed to meet, — unhappy also because of the fear that he might never meet her again.

But scarcely had he passed from the gateway into the street, when he saw a young girl walking alone in the same direction that he was going; and, even in the dusk of the dawn, he recognized her at once as the person to whom he had been introduced before the temple of Benten. As he quickened his pace to overtake her, she turned and saluted him with a graceful bow. Then for the first time he ventured to speak to her; and she answered him in a voice of which the sweetness filled his heart with joy. Through the yet silent streets they walked on, chatting happily, till they found themselves before the house where Baishū lived. There he paused — spoke to the girl of his hopes and fears. Smiling, she asked: — "Do you not know that I was sent for to become your wife?" And she entered with him.

Becoming his wife, she delighted him beyond expectation by the charm of her mind and heart. Moreover, he found her to be much more accomplished than he had supposed. Besides being able to write so wonderfully, she could paint beautiful pictures; she knew the art of arranging flowers, the art of embroidery, the art of music; she could weave and sew; and she knew everything in regard to the management of a house.

It was in the early autumn that the young people had met; and they lived together in perfect accord until the winter season began. Nothing, during those months, occurred to disturb their peace. Baishū's love for his gentle wife only strengthened with the passing of time. Yet, strangely enough, he remained ignorant of her history,—knew nothing about her family. Of such matters she had never spoken; and, as the Gods had given her to him, he imagined that it would not be proper to question her. But neither the Old-Man-under-the-Moon nor any one else came—as he had feared—to take her away. Nobody even made any inquiries about her. And the neighbors, for some undiscoverable reason, acted as if totally unaware of her presence.

Baishū wondered at all this. But stranger experiences were awaiting him.

One winter morning he happened to be passing through a somewhat remote quarter of the city, when he heard himself loudly called by name, and saw a man-servant making signs to him from the gateway of a private residence. As Baishū did not know the man's face, and did not have a single acquaintance in that part of Kyōto, he was more than startled by so abrupt a summons. But the servant, coming forward, saluted him with the utmost respect, and said, "My master greatly desires the honor of speaking with you: deign to enter for a moment." After an instant of hesitation, Baishū allowed himself to be conducted to the house. A dignified and richly dressed person, who seemed to be the master, welcomed him at the entrance, and led him to the guest-room. When the courtesies due upon a first meeting had been fully exchanged, the host apologized for the informal manner of his invitation, and said: —

"It must have seemed to you very rude of us to call you in such a way. But perhaps you will pardon our impoliteness when I tell you that we acted thus upon what I firmly believe to have been an inspiration from the Goddess Benten. Now permit me to explain.

"I have a daughter, about sixteen years old, who can write rather well,[1] and do other things in the common way: she has the ordinary nature of woman. As we were anxious to make her happy by finding a good husband for her, we prayed the Goddess Benten to help us; and we sent to every temple of Benten in the city a *tanzaku* written by the girl. Some nights later, the Goddess appeared to me in a dream, and said: 'We have heard your prayer, and have already introduced your daughter to the person who is to become her husband. During the coming winter he will visit you.' As I did not understand this assurance that a presentation had been made, I felt some doubt; I thought that the dream might have been only a common dream, signifying nothing. But last night again I saw Benten-Sama in a dream; and she said to me: 'To-morrow the young man, of whom I

---

[1] As it is the old Japanese rule that parents should speak depreciatingly of their children's accomplishments the phrase "rather well" in this connection would mean, for the visitor, "wonderfully well." For the same reason the expressions "common way" and "ordinary nature," as subsequently used, would imply almost the reverse of the literal meaning.

once spoke to you, will come to this street: then you can call him into your house, and ask him to become the husband of your daughter. He is a good young man; and later in life he will obtain a much higher rank than he now holds.' Then Benten-Sama told me your name, your age, your birthplace, and described your features and dress so exactly that my servant found no difficulty in recognizing you by the indications which I was able to give him."

This explanation bewildered Baishū instead of reassuring him; and his only reply was a formal return of thanks for the honor which the master of the house had spoken of doing him. But when the host invited him to another room, for the purpose of presenting him to the young lady, his embarrassment became extreme. Yet he could not reasonably decline the introduction. He could not bring himself, under such extraordinary circumstances, to announce that he already had a wife,—a wife given to him by the Goddess Benten herself; a wife from whom he could not even think of separating. So, in silence and trepidation, he followed his host to the apartment indicated.

Then what was his amazement to discover, when presented to the daughter of the house, that she was the very same person whom he had already taken to wife!

The same,—yet not the same.

She to whom he had been introduced by the Old-Man-under-the-Moon, was only the soul of the beloved.

She to whom he was now to be wedded, in her father's house, was the body.

Benten had wrought this miracle for the sake of her worshippers.

***

The original story breaks off suddenly at this point, leaving several matters unexplained. The ending is rather unsatisfactory. One would like to know something about the mental experiences of the real maiden during the married life of her phantom. One would also like to know what became of the phantom,—whether it continued to lead an independent existence; whether it waited patiently for the return of its husband; whether it paid a visit to the real bride. And the book says nothing about these things. But a Japanese friend explains the miracle thus: —

"The spirit-bride was really formed out of the *tanzaku.* So it is possible that the real girl did not know anything about the meeting at the temple of Benten. When she wrote those beautiful characters upon the *tanzaku,* something of her spirit passed into them. Therefore it was possible to evoke from the writing the double of the writer."

The Gratitude of the Samébito[1]

[1] The original of this story may be found in the book called *Kibun-Anbaiyoshi*

The Gratitude of the Samébito

THERE was a man named Tawaraya Tōtarō, who lived in the Province of Ōmi. His house was situated on the shore of Lake Biwa, not far from the famous temple called Ishiyamadera. He had some property, and lived in comfort; but at the age of twenty-nine he was still unmarried. His greatest ambition was to marry a very beautiful woman; and he had not been able to find a girl to his liking.

One day, as he was passing over the Long Bridge of Séta,[1] he saw a strange being crouching close to the parapet. The body of this being resembled

[1] The Long Bridge of Séta (*Séta-no-Naga-Hashi*), famous in Japanese legend, is nearly eight hundred feet in length, and commands a beautiful view. This bridge crosses the waters of the Setagawa near the junction of the stream with Lake Biwa. Ishiyamadera, one of the most picturesque Buddhist temples in Japan, is situated within a short distance from the bridge.

the body of a man, but was black as ink; its face was like the face of a demon; its eyes were green as emeralds; and its beard was like the beard of a dragon. Tōtarō was at first very much startled. But the green eyes looked at him so gently that after a moment's hesitation he ventured to question the creature. Then it answered him, saying: "I am a *Samébito,*[1]—a Shark-Man of the sea; and until a short time ago I was in the service of the Eight Great Dragon-Kings [*Hachi-Dai-Ryū-Ō*] as a subordinate officer in the Dragon-Palace [*Ryūgū*].[2] But because of a small fault which I committed, I was dismissed from the Dragon-Palace, and also banished from the Sea. Since then I have been wandering about here,—unable to get any food, or even a place to lie down. If you can

---

[1] Literally, "a Shark-Person," but in this story the *Samébito* is a male. The characters for *Samébito* can also be read *Kōjin,*—which is the usual reading. In dictionaries the word is loosely rendered by "merman" or "mermaid;" but as the above description shows, the *Samébito* or *Kōjin* of the Far East is a conception having little in common with the Western idea of a merman or mermaid.

[2] *Ryūgū* is also the name given to the whole of that fairy-realm beneath the sea which figures in so many Japanese legends.

feel any pity for me, do, I beseech you, help me to find a shelter, and let me have something to eat!"

This petition was uttered in so plaintive a tone, and in so humble a manner, that Tōtarō's heart was touched. "Come with me," he said. "There is in my garden a large and deep pond where you may live as long as you wish; and I will give you plenty to eat."

The *Samébito* followed Tōtarō home, and appeared to be much pleased with the pond.

Thereafter, for nearly half a year, this strange guest dwelt in the pond, and was every day supplied by Tōtarō with such food as sea-creatures like.

[*From this point of the original narrative the Shark-Man is referred to, not as a monster, but as a sympathetic Person of the male sex.*]

Now, in the seventh month of the same year, there was a female pilgrimage (*nyonin-mōdé*) to the great Buddhist temple called Miidera, in the neighboring town of Ōtsu; and Tōtarō went to Ōtsu to attend the festival. Among the multitude of women and young girls there assembled, he observed a person of extraordinary beauty. She seemed about sixteen years old; her face was fair and pure as snow; and the loveliness of her lips assured the beholder that their every utterance would sound "as sweet as the voice of a nightingale singing upon a plum-tree." Tōtarō fell in love with her at sight. When she left the temple he followed her at a respectful distance, and discovered that she and her mother were staying for a few days at a certain house in the neighboring village of Séta. By questioning some of the village folk, he was able also to learn that her name was Tamana; that she was unmarried; and that her family appeared to be unwilling that she should marry a man of ordinary rank,—for they demanded as a betrothal-gift a casket containing ten thousand jewels.[1]

Tōtarō returned home very much dismayed by this information. The more that he thought about the strange betrothal-gift demanded by the girl's parents, the more he felt that he could never

---

[1] *Tama* in the original. This word *tama* has a multitude of meanings; and as here used it is quite as indefinite as our own terms "jewel," "gem," or "precious stone." Indeed, it is more indefinite, for it signifies also a bead of coral, a ball of crystal, a polished stone attached to a hairpin, etc., etc. Later on, however, I venture to render it by "ruby,"—for reasons which need no explanation.

expect to obtain her for his wife. Even supposing that there were as many as ten thousand jewels in the whole country, only a great prince could hope to procure them.

But not even for a single hour could Tōtarō banish from his mind the memory of that beautiful being. It haunted him so that he could neither eat nor sleep; and it seemed to become more and more vivid as the days went by. And at last he became ill,—so ill that he could not lift his head from the pillow. Then he sent for a doctor.

The doctor, after having made a careful examination, uttered an exclamation of surprise. "Almost any kind of sickness," he said, "can be cured by proper medical treatment, except the sickness of love. Your ailment is evidently love-sickness. There is no cure for it. In ancient times Rōya-Ō Hakuyo died of that sickness; and you must prepare yourself to die as he died." So saying, the doctor went away, without even giving any medicine to Tōtarō.

About this time the Shark-Man that was living in the garden-pond heard of his master's sickness, and came into the house to wait upon Tōtarō.

And he tended him with the utmost affection both by day and by night. But he did not know either the cause or the serious nature of the sickness until nearly a week later, when Tōtarō, thinking himself about to die, uttered these words of farewell: —

"I suppose that I have had the pleasure of caring for you thus long, because of some relation that grew up between us in a former state of existence. But now I am very sick indeed, and every day my sickness becomes worse; and my life is like the morning dew which passes away before the setting of the sun. For your sake, therefore, I am troubled in mind. Your existence has depended upon my care; and I fear that there will be no one to care for you and to feed you when I am dead. . . . My poor friend! . . . Alas! our hopes and our wishes are always disappointed in this unhappy world!"

No sooner had Tōtarō spoken these words than the Samébito uttered a strange wild cry of pain, and began to weep bitterly. And as he wept, great tears of blood streamed from his green eyes and rolled down his black cheeks and dripped upon the floor. And, falling, they were blood; but, having fallen, they became hard and bright and beautiful, — became jewels of inestimable price, rubies splendid as crimson fire. For when men of the sea weep, their tears become precious stones.

Then Tōtarō, beholding this marvel, was so amazed and overjoyed that his strength returned to him. He sprang from his bed, and began to pick up and to count the tears of the Shark-Man, crying out the while: "My sickness is cured! I shall live! I shall live!"

Therewith, the Shark-Man, greatly astonished, ceased to weep, and asked Tōtarō to explain this wonderful cure; and Tōtarō told him about the young person seen at Miidera, and about the extraordinary marriage-gift demanded by her family. "As I felt sure," added Tōtarō, "that I should never be able to get ten thousand jewels, I supposed that my suit would be hopeless. Then I became very unhappy, and at last fell sick. But now, because of your generous weeping, I have many precious stones; and I think that I shall be able to marry that girl. Only — there are not yet quite enough stones; and I beg that you will be good enough to weep a little more, so as to make up the full number required."

But at this request the Samébito shook his head, and answered in a tone of surprise and of reproach : —

" Do you think that I am like a harlot, — able to weep whenever I wish ? Oh, no ! Harlots shed tears in order to deceive men ; but creatures of the sea cannot weep without feeling real sorrow. I wept for you because of the true grief that I felt in my heart at the thought that you were going to die. But now I cannot weep for you, because you have told me that your sickness is cured."

" Then what am I to do ? " plaintively asked Tōtarō. " Unless I can get ten thousand jewels, I cannot marry the girl ! "

The Samébito remained for a little while silent, as if thinking. Then he said : —

" Listen ! To-day I cannot possibly weep any more. But to-morrow let us go together to the Long Bridge of Séta, taking with us some wine and some fish. We can rest for a time on the bridge ; and while we are drinking the wine and eating the fish, I shall gaze in the direction of the Dragon-Palace, and try, by thinking of the happy days that I spent there, to make myself feel homesick — so that I can weep."

Tōtarō joyfully assented.

Next morning the two, taking plenty of wine and fish with them, went to the Séta bridge, and rested there, and feasted. After having drunk a great deal of wine, the Samébito began to gaze in the direction of the Dragon-Kingdom, and to think about the past. And gradually, under the softening influence of the wine, the memory of happier days filled his heart with sorrow, and the pain of homesickness came upon him, so that he could weep profusely. And the great red tears that he shed fell upon the bridge in a shower of rubies; and Tōtarō gathered them as they fell, and put them into a casket, and counted them until he had counted the full number of ten thousand. Then he uttered a shout of joy.

Almost in the same moment, from far away over the lake, a delightful sound of music was heard; and there appeared in the offing, slowly rising from the waters, like some fabric of cloud, a palace of the color of the setting sun.

At once the Samébito sprang upon the parapet of the bridge, and looked, and laughed for joy. Then, turning to Tōtarō, he said: —

"There must have been a general amnesty proclaimed in the Dragon-Realm; the Kings are calling me. So now I must bid you farewell.

I am happy to have had one chance of befriending you in return for your goodness to me."

With these words he leaped from the bridge; and no man ever saw him again. But Tōtarō presented the casket of red jewels to the parents of Tamana, and so obtained her in marriage.

JAPANESE STUDIES

. . . Life ere long
Came on me in the public ways, and bent
Eyes deeper than of old: Death met I too,
And saw the dawn glow through.

—GEORGE MEREDITH

Sémi
(*CICADÆ*)

Koë ni mina
Naki-shimōté ya—
Sémi no kara!
—*Japanese Love-Song*

The voice having been all consumed by crying, there remains only the shell of the *sémi!*

Sémi

I

A CELEBRATED Chinese scholar, known in Japanese literature as Riku-Un, wrote the following quaint account of the Five Virtues of the Cicada: —

"I. — The Cicada has upon its head certain figures or signs.[1] These represent its [written] characters, style, literature.

"II. — It eats nothing belonging to earth, and drinks only dew. This proves its cleanliness, purity, propriety.

"III. — It always appears at a certain fixed time. This proves its fidelity, sincerity, truthfulness.

"IV. — It will not accept wheat or rice. This proves its probity, uprightness, honesty.

---

1 The curious markings on the head of one variety of Japanese *sémi* are believed to be characters which are names of souls.

"V.—It does not make for itself any nest to live in. This proves its frugality, thrift, economy."

We might compare this with the beautiful address of Anacreon to the cicada, written twenty-four hundred years ago: on more than one point the Greek poet and the Chinese sage are in perfect accord:—

"*We deem thee happy, O Cicada, because, having drunk, like a king, only a little dew, thou dost chirrup on the tops of trees. For all things whatsoever that thou seest in the fields are thine, and whatsoever the seasons bring forth. Yet art thou the friend of the tillers of the land,—from no one harmfully taking aught. By mortals thou art held in honor as the pleasant harbinger of summer; and the Muses love thee. Phœbus himself loves thee, and has given thee a shrill song. And old age does not consume thee. O thou gifted one,—earth-born, song-loving, free from pain, having flesh without blood,—thou art nearly equal to the Gods!*"[1]

[1] In this and other citations from the Greek anthology, I have depended upon Burges' translation.

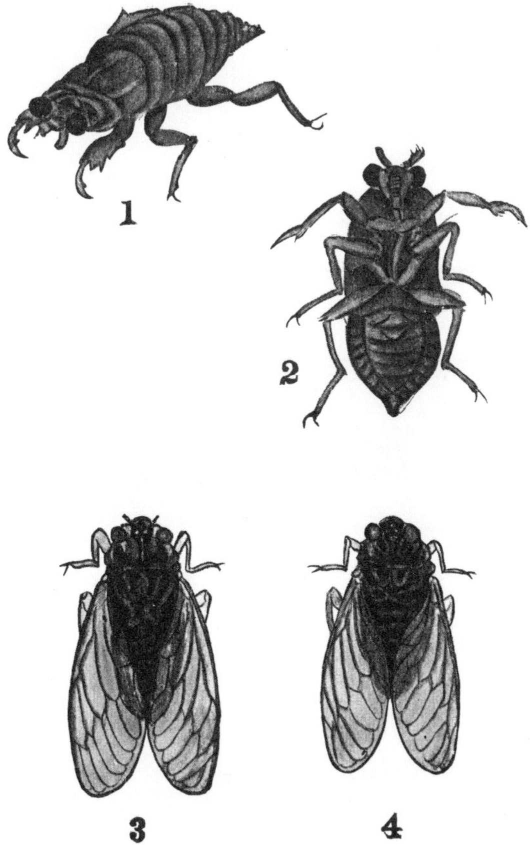

PLATE I.

1–2, *Young Sémi.*
3–4, *Haru–Zémi*, also called *Nawashiro–Zémi.*

And we must certainly go back to the old Greek literature in order to find a poetry comparable to that of the Japanese on the subject of musical insects. Perhaps of Greek verses on the cricket, the most beautiful are the lines of Meleager: "*O cricket, the soother of slumber . . . weaving the thread of a voice that causes love to wander away!*" . . . There are Japanese poems scarcely less delicate in sentiment on the chirruping of night-crickets; and Meleager's promise to reward the little singer with gifts of fresh leek, and with "drops of dew cut up small," sounds strangely Japanese. Then the poem attributed to Anyté, about the little girl Myro making a tomb for her pet cicada and cricket, and weeping because Hades, "hard to be persuaded," had taken her playthings away, represents an experience familiar to Japanese child-life. I suppose that little Myro — (how freshly her tears still glisten, after seven and twenty centuries!) — prepared that "common tomb" for her pets much as the little maid of Nippon would do to-day, putting a small stone on top to serve for a monument. But the wiser Japanese Myro would repeat over the grave a certain Buddhist prayer.

It is especially in their poems upon the cicada that we find the old Greeks confessing their love of insect-melody: witness the lines in the Anthology about the tettix caught in a spider's snare, and "making lament in the thin fetters" until freed by the poet;—and the verses by Leonidas of Tarentum picturing the "unpaid minstrel to wayfaring men" as "sitting upon lofty trees, warmed with the great heat of summer, sipping the dew that is like woman's milk;"—and the dainty fragment of Meleager, beginning: "*Thou vocal tettix, drunk with drops of dew, sitting with thy serrated limbs upon the tops of petals, thou givest out the melody of the lyre from thy dusky skin.*" . . . Or take the charming address of Evenus to a nightingale:—

"*Thou Attic maiden, honey-fed, hast chirping seized a chirping cicada, and bearest it to thy unfledged young,—thou, a twitterer, the twitterer,—thou, the winged, the well-winged,—thou, a stranger, the stranger,—thou, a summer-child, the summer-child! Wilt thou not quickly cast it from thee? For it is not right, it is not just, that those engaged in song should perish by the mouths of those engaged in song.*"

On the other hand, we find Japanese poets much more inclined to praise the voices of night-crickets than those of sémi. There are countless poems about sémi, but very few which commend their singing. Of course the sémi are very different from the cicadæ known to the Greeks. Some varieties are truly musical; but the majority are astonishingly noisy, — so noisy that their stridulation is considered one of the great afflictions of summer. Therefore it were vain to seek among the myriads of Japanese verses on sémi for anything comparable to the lines of Evenus above quoted; indeed, the only Japanese poem that I could find on the subject of a cicada caught by a bird, was the following: —

Ana kanashi!
Tobi ni toraruru
Sémi no koë.
— RANSETSU.

Ah! how piteous the cry of the sémi seized by the kite!

Or "caught by a boy" the poet might equally well have observed, — this being a much more frequent cause of the pitiful cry. The lament of

Nicias for the tettix would serve as the elegy of many a sémi:—

"*No more shall I delight myself by sending out a sound from my quick-moving wings, because I have fallen into the savage hand of a boy, who seized me unexpectedly, as I was sitting under the green leaves.*"

Here I may remark that Japanese children usually capture sémi by means of a long slender bamboo tipped with bird-lime (*mochi*). The sound made by some kinds of sémi when caught is really pitiful,—quite as pitiful as the twitter of a terrified bird. One finds it difficult to persuade oneself that the noise is not a *voice* of anguish, in the human sense of the word "voice," but the production of a specialized exterior membrane. Recently, on hearing a captured sémi thus scream, I became convinced in quite a new way that the stridulatory apparatus of certain insects must not be thought of as a kind of musical instrument, but as an organ of speech, and that its utterances are as intimately associated with simple forms of emotion, as are the notes of a bird,—the extraordinary difference being that the insect has its vocal chords *outside.*

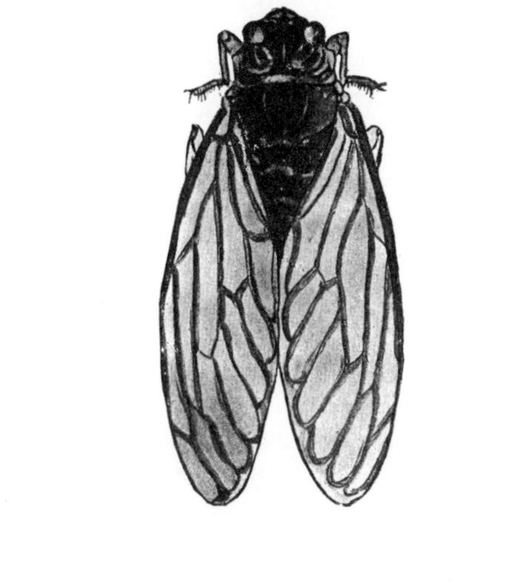

PLATE II.

"*Shinnó Shinnó*"
Also called *Yama-Zémi*, and *Kuma-Zémi*

But the insect-world is altogether a world of goblins and fairies: creatures with organs of which we cannot discover the use, and senses of which we cannot imagine the nature; — creatures with myriads of eyes, or with eyes in their backs, or with eyes moving about at the ends of trunks and horns; — creatures with ears in their legs and bellies, or with brains in their waists! If some of them happen to have voices outside of their bodies instead of inside, the fact ought not to surprise anybody.

I have not yet succeeded in finding any Japanese verses alluding to the stridulatory apparatus of sémi, — though I think it probable that such verses exist. Certainly the Japanese have been for centuries familiar with the peculiarities of their own singing insects. But I should not now presume to say that their poets are incorrect in speaking of the "voices" of crickets and of cicadæ. The old Greek poets who actually describe insects as producing music with their wings and feet, nevertheless speak of the "voices," the "songs," and the "chirruping" of such creatures, — just as the Japanese poets do. For example, Meleager thus addresses the cricket:

"O thou that art with shrill wings the self-formed imitation of the lyre, chirrup me something pleasant while beating your vocal wings with your feet! . . ."

II

BEFORE speaking further of the poetical literature of sémi, I must attempt a few remarks about the sémi themselves. But the reader need not expect anything entomological. Excepting, perhaps, the butterflies, the insects of Japan are still little known to men of science; and all that I can say about sémi has been learned from inquiry, from personal observation, and from old Japanese books of an interesting but totally unscientific kind. Not only do the authors contradict each other as to the names and characteristics of the best-known sémi; they attach the word sémi to names of insects which are not cicadæ.

The following enumeration of sémi is certainly incomplete; but I believe that it includes the better-known varieties and the best melodists. I must ask the reader, however, to bear in mind that the time of the appearance of certain sémi differs in different parts of Japan; that the same kind of sémi may be called by different names in different provinces; and that these notes have been written in Tōkyō.

I. — HARU-ZÉMI.

VARIOUS small sémi appear in the spring. But the first of the big sémi to make itself heard is the *haru-zémi* ("spring-sémi"), also called *uma-zémi* ("horse-sémi"), *kuma-zémi* ("bear-sémi"), and other names. It makes a shrill wheezing sound, — *ji-i-i-i-i-iiiiiii*, — beginning low, and gradually rising to a pitch of painful intensity. No other cicada is so noisy as the *haru-zémi;* but the life of the creature appears to end with the season. Probably this is the sémi referred to in an old Japanese poem: —

Hatsu-sémi ya!
"Koré wa atsui" to
Iu hi yori.
— TAIMU.

The day after the first day on which we exclaim, "Oh, how hot it is!" the first sémi begins to cry.

II. — "SHINNÉ-SHINNÉ."

THE *shinnė-shinnė* — also called *yama-zėmi*, or "mountain-sémi"; *kuma-zémi*, or "bear-sémi";

and *ō-sèmi*, or " great sémi " — begins to sing as early as May. It is a very large insect. The upper part of the body is almost black, and the belly a silvery-white; the head has curious red markings. The name *shinnè-shinnè* is derived from the note of the creature, which resembles a quick continual repetition of the syllables *shinnè*. About Kyōto this sémi is common: it is rarely heard in Tōkyō.

[My first opportunity to examine an *ō-sèmi* was in Shidzuoka. Its utterance is much more complex than the Japanese onomatope implies; I should liken it to the noise of a sewing-machine in full operation. There is a double sound: you hear not only the succession of sharp metallic clickings, but also, below these, a slower series of dull clanking tones. The stridulatory organs are light green, looking almost like a pair of tiny green leaves attached to the thorax.]

III. — ABURAZÉMI.

THE *aburazèmi*, or "oil-sémi," makes its appearance early in the summer. I am told that it owes its name to the fact that its shrilling resembles the sound of oil or grease frying in a pan.

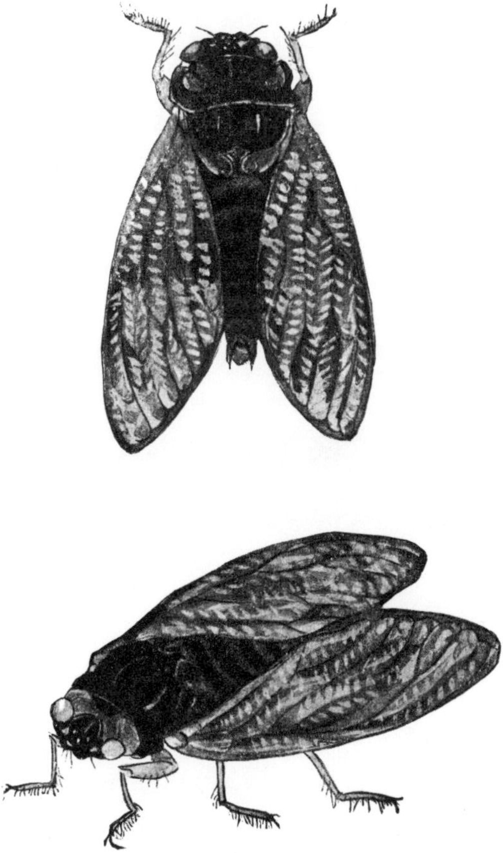

PLATE III.

Aburazémi.

Some writers say that the shrilling resembles the sound of the syllables *gacharin-gacharin;* but others compare it to the noise of water boiling. The *aburazémi* begins to chant about sunrise; then a great soft hissing seems to ascend from all the trees. At such an hour, when the foliage of woods and gardens still sparkles with dew, might have been composed the following verse, — the only one in my collection relating to the *aburazémi:* —

Ano koë dé
Tsuyu ga inochi ka? —
Aburazémi!

Speaking with that voice, has the dew taken life? — Only the *aburazémi!*

IV. — MUGI-KARI-ZÉMI.

THE *mugi-kari-zémi*, or "barley-harvest sémi," also called *goshiki-zémi*, or "five-colored sémi," appears early in the summer. It makes two distinct sounds in different keys, resembling the syllables *shi-in, shin — chi-i, chi-i.*

V. — HIGURASHI, OR "KANA-KANA."

THIS insect, whose name signifies "day-darkening," is the most remarkable of all the Japanese cicadæ. It is not the finest singer among them; but even as a melodist it ranks second only to the *tsuku-tsuku-bōshi*. It is the special minstrel of twilight, singing only at dawn and sunset; whereas most of the other sémi make their music only in the full blaze of day, pausing even when rain-clouds obscure the sun. In Tōkyō the *higurashi* usually appears about the end of June, or the beginning of July. Its wonderful cry, — *kana-kana-kana-kana-kana*, — beginning always in a very high clear key, and slowly descending, is almost exactly like the sound of a good hand-bell, very quickly rung. It is not a clashing sound, as of violent ringing; it is quick, steady, and of surprising sonority. I believe that a single *higurashi* can be plainly heard a quarter of a mile away; yet, as the old Japanese poet Yayū observed, "no matter how many *higurashi* be singing together, we never find them noisy." Though powerful and penetrating as a resonance of metal, the *higurashi's* call is musical even to the degree of sweetness; and there is a peculiar melancholy in it that accords with the hour of gloaming. But the most astonishing fact in regard to the cry of the *higurashi* is the individual quality characterizing the note of each insect.

No two *higurashi* sing precisely in the same tone. If you hear a dozen of them singing at once, you will find that the timbre of each voice is recognizably different from every other. Certain notes ring like silver, others vibrate like bronze; and, besides varieties of timbre suggesting bells of various weight and composition, there are even differences in tone, that suggest different *forms* of bell.

I have already said that the name *higurashi* means "day-darkening," — in the sense of twilight, gloaming, dusk; and there are many Japanese verses containing plays on the word, — the poets affecting to believe, as in the following example, that the crying of the insect hastens the coming of darkness: —

Higurashi ya!
Sutétéoitémo
Kururu hi wo.

O Higurashi! — even if you let it alone, day darkens fast enough!

This, intended to express a melancholy mood, may seem to the Western reader far-fetched. But another little poem — referring to the effect of the sound upon the conscience of an idler — will be appreciated by any one accustomed to hear the *higurashi*. I may observe, in this connection, that the first clear evening cry of the insect is quite as startling as the sudden ringing of a bell: —

Higurashi ya!
Kyō no kétai wo
Omou-toki.
—Rikei.

Already, O Higurashi, your call announces the evening!
Alas, for the passing day, with its duties left undone!

VI. — "Minmin"–Zémi.

The *minmin-zémi* begins to sing in the Period of Greatest Heat. It is called "*min-min*" because its note is thought to resemble the syllable "*min*" repeated over and over again, — slowly at first, and very loudly; then more and more quickly and softly, till the utterance dies away in a sort of buzz: "*min — min — min-min-min-minminmin-dzzzzzzz.*" The sound is plaintive, and not unpleasing. It is often compared to the sound of the voice of a priest chanting the *sûtras*.

VII. — Tsuku-tsuku-bōshi.

On the day immediately following the Festival of the Dead, by the old Japanese calendar[1]

[1] That is to say, upon the 16th day of the 7th month.

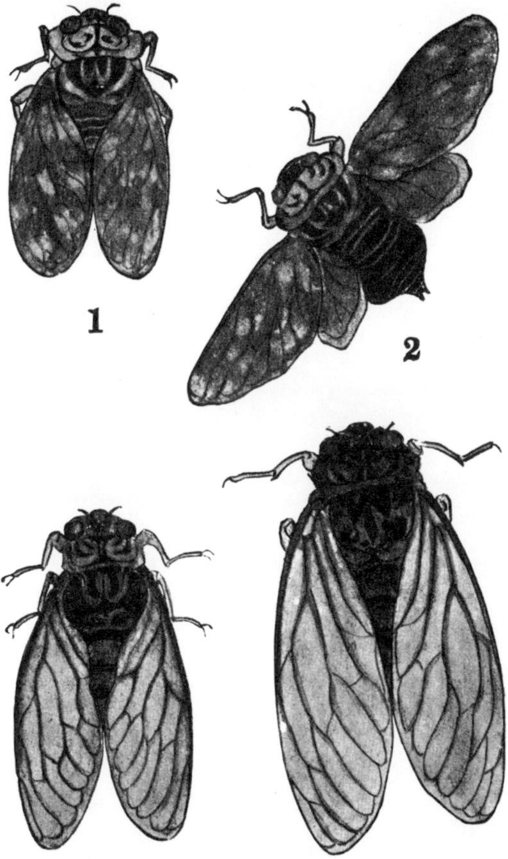

PLATE IV.

1–2, *Mugikari–Zémi*, also called *Goshiki–Zémi.*
3, *Higurashi.*
4, "*Min–Min–Zémi.*"

(which is incomparably more exact than our Western calendar in regard to nature-changes and manifestations), begins to sing the *tsuku-tsuku-bōshi.* This creature may be said to sing like a bird. It is also called *kutsu-kutsu-bōshi*, *chōko-chōko-uisu*, *tsuku-tsuku-hōshi*, *tsuku-tsuku-oïshi*, — all onomatopoetic appellations. The sounds of its song have been imitated in different ways by various writers. In Izumo the common version is, —

Tsuku-tsuku-uisu,
Tsuku-tsuku-uisu,
Tsuku-tsuku-uisu: —
  Ui-ōsu
  Ui-ōsu
  Ui-ōsu
  Ui-ōs-s-s-s-s-s-s-s-su.

Another version runs, —

Tsuku-tsuku-uisu,
Tsuku-tsuku-uisu,
Tsuku-tsuku-uisu: —
  Chi-i yara!
  Chi-i yara!
  Chi-i yara!
  Chi-i, chi, chi, chi, chi, chiii.

But some say that the sound is *Tsukushi-koïshi.* There is a legend that in old times a man of Tsukushi (the ancient name of Kyūshū) fell sick and died while far away from home, and that the ghost of him became an autumn cicada, which cries unceasingly, *Tsukushi-koïshi!* — *Tsukushi-koïshi!* ("I long for Tsukushi! — I want to see Tsukushi!")

It is a curious fact that the earlier sémi have the harshest and simplest notes. The musical sémi do not appear until summer; and the *tsuku-tsuku-bōshi*, having the most complex and melodious utterance of all, is one of the latest to mature.

VIII. — Tsurigané-Sémi.[1]

The *tsuriganė-sémi* is an autumn cicada. The word *tsuriganė* means a suspended bell, — especially the big bell of a Buddhist temple. I am somewhat puzzled by the name; for the insect's music really suggests the tones of a Japanese harp, or *koto* — as good authorities declare. Perhaps the appellation refers not to the boom of the bell, but to those deep, sweet hummings which follow after the peal, wave upon wave.

---

[1] This sémi appears to be chiefly known in Shikoku.

III

JAPANESE poems on sémi are usually very brief; and my collection chiefly consists of *hokku*,—compositions of seventeen syllables. Most of these *hokku* relate to the sound made by the sémi,—or, rather, to the sensation which the sound produced within the poet's mind. The names attached to the following examples are nearly all names of old-time poets,—not the real names, of course, but the *gō*, or literary names by which artists and men of letters are usually known.

Yokoi Yayū, a Japanese poet of the eighteenth century, celebrated as a composer of *hokku*, has left us this naïve record of the feelings with which he heard the chirruping of cicadæ in summer and in autumn:—

"In the sultry period, feeling oppressed by the greatness of the heat, I made this verse:—

"Sémi atsushi
Matsu kirabaya to
Omou-madé.

[The chirruping of the sémi aggravates the heat until I wish to cut down the pine-tree on which it sings.]

"But the days passed quickly; and later, when I heard the crying of the sémi grow fainter and fainter in the time of the autumn winds, I began to feel compassion for them, and I made this second verse:—

"Shini-nokoré
Hitotsu bakari wa
Aki no sémi."

[Now there survives
But a single one
Of the sémi of autumn!]

Lovers of Pierre Loti (the world's greatest prose-writer) may remember in *Madame Chrysanthème* a delightful passage about a Japanese house,—describing the old dry woodwork as impregnated with sonority by the shrilling crickets of a hundred summers.[1] There is a Japanese poem containing a fancy not altogether dissimilar:—

---

[1] Speaking of his own attempt to make a drawing of the interior, he observes: "Il manque à ce logis dessiné son air frêle et sa sonorité de violon sec. Dans les traits de crayon qui représentent les boiseries, il n'y a pas la précision minutieuse avec laquelle elles sont ouvragées, ni leur antiquité extrême, ni leur propreté parfaite, *ni les vibrations de cigales qu'elles semblent avoir emmagasinées pendant des centaines d'étés dans leurs fibres desséchées.*"

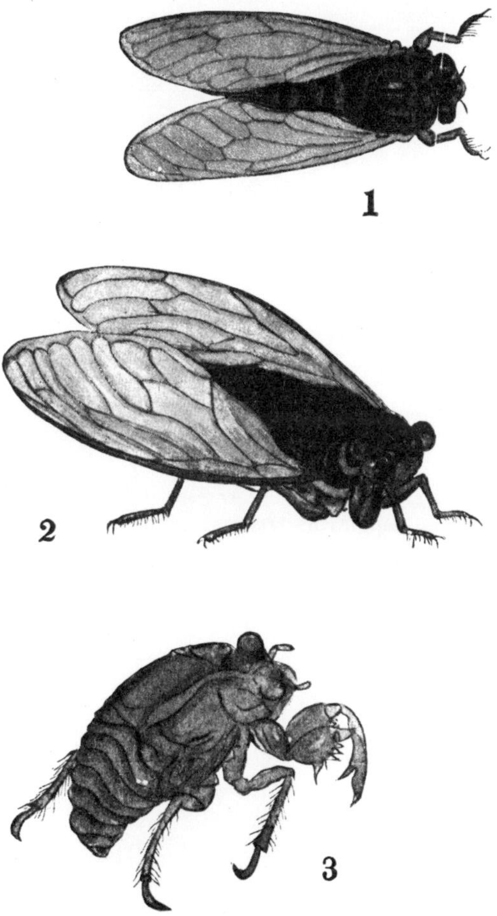

PLATE V.

1, "*Tsuku-tsuku-Bōshi*," also called "*Kutsu-kutsu-Bōshi*," etc. (*Cosmopsaltria Opalifera ?*)

2, *Tsurigané-Zémi*.

3, *The Phantom*.

Matsu no ki ni
Shimikomu gotoshi
Sémi no koë.

Into the wood of the pine-tree
Seems to soak
The voice of the sémi.

A very large number of Japanese poems about sémi describe the noise of the creatures as an affliction. To fully sympathize with the complaints of the poets, one must have heard certain varieties of Japanese cicadæ in full midsummer chorus; but even by readers without experience of the clamor, the following verses will probably be found suggestive:—

Waré hitori
Atsui yō nari,—
Sémi no koë!

—Bunsō.

Meseems that only I,—I alone among mortals,—
Ever suffered such heat!—oh, the noise of the sémi!

Ushiro kara
Tsukamu yō nari,—
Sémi no koë.

—Jofū.

Oh, the noise of the sémi!—a pain of invisible seizure,—
Clutched in an enemy's grasp,—caught by the hair from behind!

Yama no Kami no
Mimi no yamai ka?—
Sémi no koë!

—TEIKOKU.

What ails the divinity's ears?—how can the God of the Mountain
Suffer such noise to exist?—oh, the tumult of sémi!

Soko no nai
Atsusa ya kumo ni
Sémi no koë!

—SAREN.

Fathomless deepens the heat: the ceaseless shrilling of sémi
Mounts, like a hissing of fire, up to the motionless clouds.

Mizu karété,
Sémi wo fudan-no
Taki no koë.

—GEN-U.

Water never a drop: the chorus of sémi, incessant,
Mocks the tumultuous 'hiss,—the rush and foaming of rapids.

Kagéroishi
Kumo mata satté,
Sémi no koë.

—KITŌ.

Gone, the shadowing clouds!—again the shrilling of sémi
Rises and slowly swells,—ever increasing the heat!

Daita ki wa,
Ha mo ugokasazu,—
Sémi no koë!

—KAFŪ.

Somewhere fast to the bark he clung; but I cannot see him:
He stirs not even a leaf—oh! the noise of that sémi!

Tonari kara
Kono ki nikumu ya!
Sémi no koë.

— GYUKAKU.

All because of the sémi that sit and shrill on its branches—
Oh! how this tree of mine is hated now by my neighbor!

This reminds one of Yayū. We find another poet compassionating a tree frequented by sémi:—

Kazé wa mina
Sémi ni suwarété,
Hito-ki kana!

— CHŌSUI.

Alas! poor solitary tree!—pitiful now your lot,—every breath of air having been sucked up by the sémi!

Sometimes the noise of the sémi is described as a moving force:—

Sémi no koë
Ki-gi ni ugoité,
Kazé mo nashi!

— SŌYŌ.

Every tree in the wood quivers with clamor of sémi:
Motion only of noise—never a breath of wind!

Také ni kité,
Yuki yori omoshi
Sémi no koë.

— TŌGETSU.

More heavy than winter-snow the voices of perching sémi:
See how the bamboos bend under the weight of their song![1]

Morogoë ni
Yama ya ugokasu,
Ki-gi no sémi.

All shrilling together, the multitudinous sémi
Make, with their ceaseless clamor, even the mountain move.

Kusunoki mo
Ugoku yō nari,
Sémi no koë.
— Baijaku.

Even the camphor-tree seems to quake with the clamor of sémi!

Sometimes the sound is compared to the noise of boiling water: —

Hizakari wa
Niétatsu sémi no
Hayashi kana!

In the hour of heaviest heat, how simmers the forest with sémi!

Niété iru
Mizu bakari nari —
Sémi no koë.
— Taimu.

---

[1] Japanese artists have found many a charming inspiration in the spectacle of bamboos bending under the weight of snow clinging to their tops.

Simmers all the air with sibilation of sémi,
Ceaseless, wearying sense, — a sound of perpetual boiling.

Other poets complain especially of the multitude of the noise-makers and the ubiquity of the noise: —

Aritaké no
Ki ni hibiki-kéri
Sémi no koë.

How many soever the trees, in each rings the voice of the sémi.

Matsubara wo
Ichi ri wa kitari,
Sémi no koë.
— SENGA.

Alone I walked for miles into the wood of pine-trees:
Always the one same sémi shrilled its call in my ears.

Occasionally the subject is treated with comic exaggeration: —

Naité iru
Ki yori mo futoshi
Sémi no koë.

The voice of the sémi is bigger [*thicker*] than the tree on which it sings.

Sugi takashi
Sarédomo sémi no
Amaru koë!

High though the cedar be, the voice of the sémi is incomparably higher!

Koë nagaki
Sémi wa mijikaki
Inochi kana!

How long, alas! the voice and how short the life of the sémi!

Some poets celebrate the negative form of pleasure following upon the cessation of the sound:—

Sémi ni dété,
Hotaru ni modoru,—
Suzumi kana!
—Yayū.

When the sémi cease their noise, and the fireflies come out—oh! how refreshing the hour!

Sémi no tatsu,
Ato suzushisa yo!
Matsu no koë.
—Baijaku.

When the sémi cease their storm, oh, how refreshing the stillness!
Gratefully then resounds the musical speech of the pines.

[Here I may mention, by the way, that there is a little Japanese song about the *matsu no koë,* in which the onomatope "zazanza" very well represents the deep humming of the wind in the pine-needles:—

Zazanza!
Hama-matsu no oto wa,—
Zazanza,
Zazanza!

Zazanza!
The sound of the pines of the shore,—
Zazanza!
Zazanza!]

There are poets, however, who declare that the feeling produced by the noise of sémi depends altogether upon the nervous condition of the listener:—

Mori no sémi
Suzushiki koë ya,
Atsuki koë.
—OTSUSHU.

Sometimes sultry the sound; sometimes, again, refreshing:
The chant of the forest-sémi accords with the hearer's mood.

Suzushisa mo
Atsusa mo sémi no
Tokoro kana!
—FUHAKU.

Sometimes we think it cool,—the resting-place of the sémi;—sometimes we think it hot (it is all a matter of fancy).

Suzushii to
Omoéba, suzushi
Sémi no koë.
—GINKŌ.

If we think it is cool, then the voice of the sémi is cool (that is, the fancy changes the feeling).

In view of the many complaints of Japanese poets about the noisiness of sémi, the reader may be surprised to learn that out of sémi-skins there used to be made in both China and Japan — perhaps upon homœopathic principles — a medicine for the cure of ear-ache!

One poem, nevertheless, proves that sémi-music has its admirers: —

Omoshiroi zo ya,
Waga-ko no koë wa
Takai mori-ki no
　　Sémi no koë![1]

Sweet to the ear is the voice of one's own child as the voice of a sémi perched on a tall forest tree.

But such admiration is rare. More frequently the sémi is represented as crying for its nightly repast of dew: —

[1] There is another version of this poem: —

Omoshiroi zo ya,
Waga-ko no naku wa
Sembu-ségaki no
　　Kyō yori mo!

"More sweetly sounds the crying of one's own child than even the chanting of the sûtra in the service for the dead." The Buddhist service alluded to is held to be particularly beautiful.

Sémi wo kiké,—
Ichi-nichi naité
Yoru no tsuyu.
—KIKAKU.

Hear the sémi shrill! So, from earliest dawning,
All the summer day he cries for the dew of night.

Yū-tsuyu no
Kuchi ni iru madé
Naku sémi ka?
—BAISHITSU.

Will the sémi continue to cry till the night-dew fills its mouth?

Occasionally the sémi is mentioned in love-songs of which the following is a fair specimen. It belongs to that class of ditties commonly sung by geisha. Merely as a conceit, I think it pretty, in spite of the factitious pathos; but to Japanese taste it is decidedly vulgar. The allusion to beating implies jealousy:—

Nushi ni tatakaré,
Washa matsu no sémi
Sugaritsuki-tsuki
Naku bakari!

Beaten by my jealous lover,—
Like the sémi on the pine-tree
I can only cry and cling!

And indeed the following tiny picture is a truer bit of work, according to Japanese art-principles (I do not know the author's name): —

Sémi hitotsu
Matsu no yū-hi wo
Kakaé-kéri.

Lo! on the topmost pine, a solitary cicada
Vainly attempts to clasp one last red beam of sun.

IV

PHILOSOPHICAL verses do not form a numerous class of Japanese poems upon sémi; but they possess an interest altogether exotic. As the metamorphosis of the butterfly supplied to old Greek thought an emblem of the soul's ascension, so the natural history of the cicada has furnished Buddhism with similitudes and parables for the teaching of doctrine.

Man sheds his body only as the sémi sheds its skin. But each reincarnation obscures the memory of the previous one: we remember our former existence no more than the sémi remembers the shell from which it has emerged. Often a sémi may be found in the act of singing beside its cast-off skin; therefore a poet has written:—

Waré to waga
Kara ya tomurō—
Sémi no koë.
—YAYŪ.

Methinks that sémi sits and sings by his former body,—
Chanting the funeral service over his own dead self.

This cast-off skin, or simulacrum,—clinging to bole or branch as in life, and seeming still to stare with great glazed eyes,—has suggested many things both to profane and to religious poets. In love-songs it is often likened to a body consumed by passionate longing. In Buddhist poetry it becomes a symbol of earthly pomp,—the hollow show of human greatness:—

Yo no naka yo
Kaëru no hadaka,
Sémi no kinu!

Naked as frogs and weak we enter this life of trouble;
Shedding our pomps we pass: so sémi quit their skins.

But sometimes the poet compares the winged and shrilling sémi to a human ghost, and the broken shell to the body left behind:—

Tamashii wa
Ukiyo ni naité,
Sémi no kara.

Here the forsaken shell: above me the voice of the creature
Shrills like the cry of a Soul quitting this world of pain.

Then the great sun-quickened tumult of the cicadæ — landstorm of summer life foredoomed so soon to pass away — is likened by preacher and poet to the tumult of human desire. Even as the sémi rise from earth, and climb to warmth and light, and clamor, and presently again return to dust and silence, — so rise and clamor and pass the generations of men: —

Yagaté shinu
Keshiki wa miézu,
Sémi no koë.
— Bashō.

Never an intimation in all those voices of sémi
How quickly the hush will come, — how speedily all must die.

I wonder whether the thought in this little verse does not interpret something of that summer melancholy which comes to us out of nature's solitudes with the plaint of insect-voices. Unconsciously those millions of millions of tiny beings are preaching the ancient wisdom of the East, — the perpetual Sûtra of Impermanency.

Yet how few of our modern poets have given heed to the voices of insects!

Perhaps it is only to minds inexorably haunted by the Riddle of Life that Nature can speak to-day, in those thin sweet trillings, as she spake of old to Solomon.

The Wisdom of the East hears all things. And he that obtains it will hear the speech of insects, — as Sigurd, tasting the Dragon's Heart, heard suddenly the talking of birds.

NOTE. — For the pictures of sémi accompanying this paper, I am indebted to a curious manuscript work in several volumes, preserved in the Imperial Library at Uyéno. The work is entitled *Chūfu-Zusetsu*, — which might be freely rendered as "Pictures and Descriptions of Insects," — and is divided into twelve books. The writer's name is unknown; but he must have been an amiable and interesting person, to judge from the naïve preface which he wrote, apologizing for the labors of a lifetime. "When I was young," he says, "I was very fond of catching worms and insects, and making pictures of their shapes, — so that these pictures have now become several hundred in number." He believes that he has found a good reason for studying insects: "Among the multitude of living creatures in this world," he says, "those having large bodies are familiar: we know very well their names, shapes, and virtues, and the poisons which they possess. But there remain very many small creatures whose natures are still unknown, notwithstanding the fact that such little beings as insects and worms are able to injure men and to destroy what has value. So I think that it is very important for us to learn what insects or worms have special virtues or poisons." It appears that he had sent to him "from other countries" some kinds of insects "that eat the leaves and shoots of trees;" but he could not "get their exact names." For the names of domestic insects, he consulted many Chinese and Japanese books, and has been "able to write the names with the proper Chinese characters;" but he tells us that he did not fail "to pick up also the names given to worms and insects by old farmers and little boys." The preface is dated thus: — "*Ansei Kanotè, the third month — at a little cottage*" [1856].

With the introduction of scientific studies the author of the *Chūfu-Zusetsu* could no longer hope to attract attention. Yet his very modest and very beautiful work was forgotten only a moment. It is now a precious curiosity; and the old man's ghost might to-day find some happiness in a visit to the Imperial Library.

Japanese Female Names

Japanese Female Names

I

BY the Japanese a certain kind of girl is called a Rose-Girl, — *Bara-Musumé.* Perhaps my reader will think of Tennyson's "queen-rose of the rosebud-garden of girls," and imagine some analogy between the Japanese and the English idea of femininity symbolized by the rose. But there is no analogy whatever. The *Bara-Musumé* is not so called because she is delicate and sweet, nor because she blushes, nor because she is rosy; indeed, a rosy face is not admired in Japan. No; she is compared to a rose chiefly for the reason that a rose has thorns. The man who tries to pull a Japanese rose is likely to hurt his fingers. The man who tries to win a *Bara-Musumé* is apt to hurt himself much more seriously, — even unto death.

It were better, alone and unarmed, to meet a tiger than to invite the caress of a Rose-Girl.

Now the appellation of *Bara-Musumė* — much more rational as a simile than many of our own floral comparisons — can seem strange only because it is not in accord with our poetical usages and emotional habits. It is one in a thousand possible examples of the fact that Japanese similes and metaphors are not of the sort that he who runs may read. And this fact is particularly well exemplified in the *yobina*, or personal names of Japanese women. Because a *yobina* happens to be identical with the name of some tree, or bird, or flower, it does not follow that the personal appellation conveys to Japanese imagination ideas resembling those which the corresponding English word would convey, under like circumstances, to English imagination. Of the *yobina* that seem to us especially beautiful in translation, only a small number are bestowed for æsthetic reasons. Nor is it correct to suppose, as many persons still do, that Japanese girls are usually named after flowers, or graceful shrubs, or other beautiful objects. Æsthetic appellations are in use; but the majority of *yobina* are not æsthetic. Some years ago a young Japanese scholar published an interesting essay upon this subject. He had collected the personal names of about four hundred students of the Higher Normal School for Females, — girls from every part of the Empire; and he found on his list only between fifty and sixty names possessing æsthetic quality. But concerning even these he was careful to observe only that they "*caused* an æsthetic sensation," — not that they had been given for æsthetic reasons. Among them were such names as *Saki* (Cape), *Minė* (Peak), *Kishi* (Beach), *Hama* (Shore), *Kuni* (Capital), — originally place-names; — *Tsuru* (Stork), *Tazu* (Ricefield Stork), and *Chizu* (Thousand Storks); — also such appellations as *Yoshino* (Fertile Field), *Orino* (Weavers' Field), *Shirushi* (Proof), and *Masago* (Sand). Few of these could seem æsthetic to a Western mind; and probably no one of them was originally given for æsthetic reasons. Names containing the character for "Stork" are names having reference to longevity, not to beauty; and a large number of names with the termination "*no*" (field or plain) are names referring to moral qualities. I doubt whether even fifteen per cent of *yobina* are really æsthetic. A very much larger proportion are names expressing moral or mental qualities. Tenderness, kindness, deftness, cleverness, are frequently represented by *yobina ;* but appellations implying physical charm, or suggesting æsthetic ideas only, are comparatively uncommon. One reason for the fact may be that very æsthetic names are given to *geisha* and to *jōro*, and consequently vulgarized. But the chief reason certainly is that the domestic virtues still occupy in Japanese moral estimate a place not less important than that accorded to religious faith in the life of our own Middle Ages. Not in theory only, but in every-day practice, moral beauty is placed far above physical beauty; and girls are usually selected as wives, not for their good looks, but for their domestic qualities. Among the middle classes a very æsthetic name would not be considered in the best taste; among the poorer classes, it would scarcely be thought respectable. Ladies of rank, on the other hand, are privileged to bear very poetical names; yet the majority of the aristocratic *yobina* also are moral rather than æsthetic.

But the first great difficulty in the way of a study of *yobina* is the difficulty of translating them. A knowledge of spoken Japanese can help you very little indeed. A knowledge of Chinese also is indispensable. The meaning of a name written in *kana* only, — in the Japanese characters, — cannot be, in most cases, even guessed at. The Chinese characters of the name can alone explain it. The Japanese essayist, already referred to, found himself obliged to throw out no less than thirty-six names out of a list of two hundred and thirteen, simply because these thirty-six, having been recorded only in *kana*, could not be interpreted. *Kana* give only the pronunciation; and the pronunciation of a woman's name explains nothing in a majority of cases. Transliterated into Romaji, a *yobina* may signify two, three, or even half-a-dozen different things. One of the names thrown out of the list was *Banka*. *Banka* might signify "Mint" (the plant), which would be a pretty name; but it might also mean "Evening-haze." *Yuka*, another rejected name, might be an abbreviation of *Yukabutsu*, "precious"; but it might just as well mean "a floor." *Nochi*, a third example, might signify "future"; yet it could also mean "a descendant," and various other things. My reader will be able to find many other homonyms in the lists of names given further on. *Ai* in Romaji, for instance, may signify either " love " or " indigo-blue " ; — *Chō*, " a butterfly," or " superior," or " long " ; — *Ei*, either " sagacious " or " blooming " ; — *Kei*, either " rapture " or " reverence " ; — *Sato*, either " native home " or " sugar " ; — *Toshi*, either " year " or " arrowhead " ; — *Taka*, " tall," " honorable," or " falcon." The chief, and, for the present, insuperable obstacle to the use of Roman letters in writing Japanese, is the prodigious number of homonyms in the language. You need only glance into any good Japanese-English dictionary to understand the gravity of this obstacle. Not to multiply examples, I shall merely observe that there are nineteen words spelled *chō;* twenty-one spelled *ki;* twenty-five spelled *to* or *tō;* and no less than forty-nine spelled *ko* or *kō.*

Yet, as I have already suggested, the real signification of a woman's name cannot be ascertained even from a literal translation made with the help of the Chinese characters. Such a name, for instance, as *Kagami* (Mirror) really signifies the Pure-Minded, and this not in the Occidental, but in the Confucian sense of the term. *Umė*

(Plum-blossom) is a name referring to wifely devotion and virtue. *Matsu* (Pine) does not refer, as an appellation, to the beauty of the tree, but to the fact that its evergreen foliage is the emblem of vigorous age. The name *Takė* (Bamboo) is given to a child only because the bamboo has been for centuries a symbol of good-fortune. The name *Sen* (Wood-fairy) sounds charmingly to Western fancy; yet it expresses nothing more than the parents' hope of long life for their daughter and her offspring, — wood-fairies being supposed to live for thousands of years. . . . Again, many names are of so strange a sort that it is impossible to discover their meaning without questioning either the bearer or the giver; and sometimes all inquiry proves vain, because the original meaning has been long forgotten.

Before attempting to go further into the subject, I shall here offer a translation of the Tōkyō essayist's list of names, — rearranged in alphabetical order, without honorific prefixes or suffixes. Although some classes of common names are not represented, the list will serve to show the character of many still popular *yobina*, and also to illustrate several of the facts to which I have already called attention.

SELECTED NAMES OF STUDENTS AND GRADUATES OF THE HIGHER NORMAL SCHOOL FOR FEMALES (1880-1895):—

Name	Meaning	Number of students so named.
Ai	("Indigo," — the color)	1
Ai	("Love")	1
Akasuké	("The Bright Helper")	1
Asa	("Morning")	1
Asa	("Shallow")[1]	2
Au	("Meeting")	2
Bun	("Composition"—in the literary sense)[2]	1
Chika	("Near")[3]	5
Chitosé	("A Thousand Years")	1
Chiyo	("A Thousand Generations")	1
Chizu	("Thousand Storks")	1
Chō	("Butterfly")	1
Chō	("Superior")	2
Ei	("Clever")	1
Ei	("Blooming")	2
Etsu	("Delight")	1
Fudé	("Writing-brush")	1
Fuji	("Fuji," — the mountain)	1
Fuji	("Wistaria-flower")	2
Fuki	("Fuki," — name of a plant, *Nardosmia Japonica*)	1
Fuku	("Good-fortune")	2
Fumi	("Letter")[4]	5
Fumino	("Letter-field")	1

[1] Probably a place-name originally.

[2] Might we not quaintly say, "A Fair Writing"?

[3] Probably in the sense of "near and dear" — but not certainly so.

[4] *Fumi* signifies here a letter written by a woman only — a letter written according to the rules of feminine epistolary style.

Fusa . . .	("Tassel")	3
Gin . . .	("Silver")	2
Hama . .	("Shore")	3
Hana . .	("Blossom")	3
Haruë . .	("Spring-time Bay")	1
Hatsu . .	("The First-born")	2
Hidé . . .	("Excellent")	4
Hidé . . .	("Fruitful")	2
Hisano . .	("Long Plain")	2
Ichi . . .	("Market")	4
Iku . . .	("Nourishing")	3
Iné . . .	("Springing Rice")	3
Ishi . . .	("Stone")	1
Ito . . .	("Thread")	4
Iwa . . .	("Rock")	1
Jun . . .	("The Obedient")[1]	1
Kagami . .	("Mirror")	3
Kama . .	("Sickle")	1
Kamé . .	("Tortoise")	2
Kaméyo . .	("Generations-of-the-Tortoise")[2] . .	1
Kan . . .	("The Forbearing")[3]	11
Kana . .	("Character"—in the sense of written character)[4]	2
Kané . . .	("Bronze")	3

1 *Jun suru* means to be obedient unto death. The word *jun* has a much stronger signification than that which attaches to our word "obedience" in these modern times.

2 The tortoise is supposed to live for a thousand years.

3 Abbreviation of *kannin*, "forbearance," "self-control," etc. The name might equally well be translated "Patience."

4 *Kana* signifies the Japanese syllabary,—the characters with which the language is written. The reader may imagine, if he wishes, that the name signifies the Alpha and Omega of all feminine charm; but I confess that I have not been able to find any satisfactory explanation of it.

Katsu	("Victorious")	2
Kazashi	("Hair-pin,"—or any ornament worn in the hair)	1
Kazu	("Number,"—i. e., "great number")	1
Kei	("The Respectful")	3
Ken	("Humility")	1
Kiku	("Chrysanthemum")	6
Kikuë	("Chrysanthemum-branch")	1
Kikuno	("Chrysanthemum-field")	1
Kimi	("Sovereign")	1
Kin	("Gold")	4
Kinu	("Cloth-of-Silk")	1
Kishi	("Beach")	2
Kiyo	("Happy Generations")	1
Kiyo	("Pure")	5
Ko	("Chime,"—the sound of a bell)	1
Kō	("Filial Piety")	11
Kō	("The Fine")	1
Koma	("Filly")	1
Komé	("Cleaned Rice")	1
Koto	("Koto,"—the Japanese harp)	4
Kuma	("Bear")	1
Kumi	("Braid")	1
Kuni	("Capital,"—chief city)	1
Kuni	("Province")	3
Kura	("Treasure-house")	1
Kurano	("Storehouse-field")	1
Kuri	("Chestnut")	1
Kuwa	("Mulberry-tree")	1
Masa	("Straightforward,"—upright)	3
Masago	("Sand")	1
Masu	("Increase")	3
Masuë	("Branch-of-Increase")	1
Matsu	("Pine")	2

Matsuë . . ("Pine-branch") 1
Michi . . ("The Way,"—doctrine) 4
Mië . . . ("Triple Branch") 1
Mikië . . ("Main-branch") 1
Miné . . . ("Peak") 2
Mitsu . . ("Light") 5
Mitsuë . . ("Shining Branch") 1
Morië . . ("Service-Bay")[1] 1
Naka . . ("The Midmost") 4
Nami . . ("Wave") 1
Nobu . . . ("Fidelity") 6
Nobu . . . ("The Prolonger")[2] 1
Nobuë . . ("Lengthening-branch") 1
Nui . . . ("Tapestry,"—or, Embroidery) . . . 1
Orino . . ("Weaving-Field") 1
Raku . . . ("Pleasure") 3
Ren . . . ("The Arranger") 1
Riku . . . ("Land,"—ground) 1
Roku . . . ("Emolument") 1
Ryō . . . ("Dragon") 1
Ryū . . . ("Lofty") 3
Sada . . . ("The Chaste") 8
Saki . . . ("Cape,"—promontory) 1
Saku . . . ("Composition")[3] 3
Sato . . . ("Home,"—native place) 2
Sawa . . . ("Marsh") 1
Sei . . . ("Force") 1
Seki . . . ("Barrier,"—city-gate, toll-gate, etc.) . 3

---

[1] The word "service" here refers especially to attendance at meal-time,—to the serving of rice, etc.

[2] Perhaps in the hopeful meaning of extending the family-line; but more probably in the signification that a daughter's care prolongs the life of her parents, or of her husband's parents.

[3] Abbreviation of *sakubun,* a literary composition.

Sen	("Fairy")[1]	3
Setsu	("True," — tender and true)	2
Shidzu	("The Calmer")	1
Shidzu	("Peace")	2
Shigë	("Two-fold")	2
Shika	("Deer")	2
Shikaë	("Deer-Inlet")	1
Shimé	("The Clasp," — fastening)	1
Shin	("Truth")	1
Shina	("Goods")	1
Shina	("Virtue")	1
Shino	("Slender Bamboo")	1
Shirushi	("The Proof," — evidence)	1
Shun	("The Excellent")	1
Sué	("The Last")	2
Sugi	("Cedar," — cryptomeria)	1
Suté	("Forsaken," —foundling)	1
Suzu	("Little Bell")	8
Suzu	("Tin")	1
Suzuë	("Branch of Little Bells")	1
Taë	("Exquisite")	1
Taka	("Honor")	2
Taka	("Lofty")	9
Také	("Bamboo")	1
Tama	("Jewel")	1
Tamaki	("Ring")	1
Tamé	("For-the-Sake-of—")	3
Tani	("Valley")	1
Tazu	("Ricefield-Stork")	1

[1] As a matter of fact, we have no English equivalent for the word "sen," or "sennin," — signifying a being possessing magical powers of all kinds and living for thousands of years. Some authorities consider the belief in *sennin* of Indian origin, and probably derived from old traditions of the Rishi.

Tetsu . .	("Iron")	4
Toku . .	("Virtue")	2
Tomé . .	("Stop," — cease)[1]	1
Tomi . .	("Riches")	3
Tomijū . .	("Wealth-and-Longevity")	1
Tomo . .	("The Friend")	4
Tora . . .	("Tiger")	1
Toshi . .	("Arrowhead")	1
Toyo . . .	("Abundance")	3
Tsugi . .	("Next," — i. e., second in order of birth)	2
Tsuna . .	("Bond," — rope, or fetter)	1
Tsuné . .	("The Constant," — or, as we should say, Constance)	10
Tsuru . .	("Stork")	4
Umé . . .	("Plum-blossom")	1
Umégaë . .	("Plumtree-spray")	1
Uméno . .	("Plumtree-field")	2
Urano . .	("Shore-field")	1
Ushi . . .	("Cow," — or Ox)[2]	1
Uta . . .	("Poem," — or Song)	1
Wakana .	("Young *Na*," — probably the rape-plant is referred to)	1
Yaë . . .	("Eight-fold")	1
Yasu . . .	("The Tranquil")	1

[1] Such a name may signify that the parents resolved, after the birth of the girl, to have no more children.

[2] This extraordinary name is probably to be explained as a reference to date of birth. According to the old Chinese astrology, years, months, days, and hours were all named after the Signs of the Zodiac, and were supposed to have some mystic relation to those signs. I surmise that Miss Ushi was born at the Hour of the Ox, on the Day of the Ox, in the Month of the Ox and the Year of the Ox — "*Ushi no Toshi no Ushi no Tsuki no Ushi no Hi no Ushi no Koku.*"

Yō . . .	("The Positive," — as opposed to Negative or Feminine in the old Chinese philosophy; — therefore, perhaps, Masculine)	1
Yoné . . .	("Rice," — in the old sense of wealth) .	4
Yoshi . .	("The Good")	1
Yoshino . .	("Good Field")	1
Yū . . .	("The Valiant")	1
Yuri . .	("Lily")	1

It will be observed that in the above list the names referring to Constancy, Forbearance, and Filial Piety have the highest numbers attached to them.

II

A FEW of the more important rules in regard to Japanese female names must now be mentioned.

The great majority of these *yobina* are words of two syllables. Personal names of respectable women, belonging to the middle and lower classes, are nearly always dissyllables — except in cases where the name is lengthened by certain curious suffixes which I shall speak of further on. Formerly a name of three or more syllables indicated that the bearer belonged to a superior class. But, even among the upper classes to-day, female names of only two syllables are in fashion.

Among the people it is customary that a female name of two syllables should be preceded by the honorific " O," and followed by the title " San," — as *O-Matsu San*, " the Honorable Miss [or Mrs.] Pine"; *O-Umė San*, " the Honorable Miss Plum-blossom."[1] But if the name happen to have three syllables, the honorific "O" is not used. A woman named *Kikuë* ("Chrysanthemum-Branch") is not addressed as " O-Kikuë San," but only as " Kikuë San."

Before the names of ladies, the honorific " O" is no longer used as formerly, — even when the name consists of one syllable only. Instead of the prefix, an honorific suffix is appended to the *yobina*, — the suffix *ko*. A peasant girl named *Tomi* would be addressed by her equals as *O-Tomi San*. But a lady of the same name would be addressed as *Tomiko*. Mrs. Shimoda, head-teacher of the Peeresses' School, for example, has the beautiful name *Uta*. She would be addressed by letter as " Shimoda Utako," and would so sign herself in replying; — the family-

---

[1] Under certain conditions of intimacy, both prefix and title are dropped. They are dropped also by the superior in addressing an inferior; — for example, a lady would not address her maid as " *O-Yoné San*," but merely as " *Yoné*."

name, by Japanese custom, always preceding the personal name, instead of being, as with us, placed after it.

This suffix *ko* is written with the Chinese character meaning "child," and must not be confused with the word *ko,* written with a different Chinese character, and meaning "little," which so often appears in the names of dancing girls. I should venture to say that this genteel suffix has the value of a caressing diminutive, and that the name *Aiko* might be fairly well rendered by the "Amoretta" of Spenser's *Faerie Queene.* Be this as it may, a Japanese lady named *Setsu* or *Sada* would not be addressed in these days as O-Setsu or O-Sada, but as Setsuko or Sadako. On the other hand, if a woman of the people were to sign herself as Setsuko or Sadako, she would certainly be laughed at,—since the suffix would give to her appellation the meaning of "the Lady Setsu," or "the Lady Sada."

I have said that the honorific "O" is placed before the *yobina* of women of the middle and lower classes. Even the wife of a *kurumaya* would probably be referred to as the "Honorable Mrs. Such-a-one." But there are very remarkable exceptions to this general rule regarding the prefix "O." In some country-districts the common *yobina* of two syllables is made a trisyllable by the addition of a peculiar suffix; and before such trisyllabic names the "O" is never placed. For example, the girls of Wakayama, in the Province of Kii, usually have added to their *yobina* the suffix "*ë*,"[1] signifying "inlet," "bay," "frith,"—sometimes "river." Thus we find such names as *Namië* ("Wave-Bay"), *Tomië* ("Riches-Bay"), *Sumië* ("Dwelling-Bay"), *Shizuë* ("Quiet-Bay"), *Tamaë* ("Jewel-Bay"). Again there is a provincial suffix "*no*," meaning "field" or "plain," which is attached to the majority of female names in certain districts. *Yoshino* ("Fertile Field"), *Uméno* ("Plumflower Field"), *Shizuno* ("Quiet Field"), *Urano* ("Coast Field"), *Utano* ("Song Field"), are typical names of this class. A girl called *Namië* or *Kikuno* is not addressed as "O-Namië San" or "O-Kikuno San," but as "Namië San," "Kikuno San."

[1] This suffix must not be confused with the suffix "*ë*," signifying "branch," which is also attached to many popular names. Without seeing the Chinese character, you cannot decide whether the name *Tamaë*, for example, means "Jewel-branch" or "Jewel Inlet."

"San" (abbreviation of *Sama*, a word originally meaning "form," "appearance"), when placed after a female name, corresponds to either our "Miss" or "Mrs." Placed after a man's name it has at least the value of our "Mr.", — perhaps even more. The unabbreviated form *Sama* is placed after the names of high personages of either sex, and after the names of divinities: the Shintō Gods are styled the *Kami-Sama*, which might be translated as "the Lords Supreme"; the Bodhisattva Jizō is called *Jizō-Sama*, "the Lord Jizō." A lady may also be styled "Sama." A lady called *Ayako*, for instance, might very properly be addressed as Ayako Sama. But when a lady's name, independently of the suffix, consists of more than three syllables, it is customary to drop either the *ko* or the title. Thus "the Lady Ayamé" would not be spoken of as "Ayaméko Sama," but more euphoniously as "Ayamé Sama,"[1] or as "Ayaméko."

So much having been said as regards the etiquette of prefixes and suffixes, I shall now

[1] "Ayamé Sama," however, is rather familiar; and this form cannot be used by a stranger in verbal address, though a letter may be directed with the name so written. As a rule, the *ko* is the more respectful form.

attempt a classification of female names, — beginning with popular *yobina*. These will be found particularly interesting, because they reflect something of race-feeling in the matter of ethics and æsthetics, and because they serve to illustrate curious facts relating to Japanese custom. The first place I have given to names of purely moral meaning, — usually bestowed in the hope that the children will grow up worthy of them. But the lists should in no case be regarded as complete: they are only representative. Furthermore, I must confess my inability to explain the reason of many names, which proved as much of riddles to Japanese friends as to myself.

NAMES OF VIRTUES AND PROPRIETIES

O-Ai . . . . . "Love."
O-Chiē . . . . . "Intelligence."
O-Chū . . . . . "Loyalty."
O-Jin . . . . . "Tenderness," — humanity.
O-Jun . . . . . "Faithful-to-death."
O-Kaiyō . . . . "Forgiveness," — pardon.
O-Ken . . . . . "Wise," — in the sense of moral discernment.
O-Kō . . . . . "Filial Piety."
O-Masa . . . . "Righteous," — just.
O-Michi . . . . "The Way," — doctrine.
Misao . . . . . "Honor," — wifely fidelity.

O-Nao . . . . . . "The Upright," — honest.
O-Nobu . . . . . "The Faithful."
O-Rei . . . . . . "Propriety," — in the old Chinese sense.
O-Retsu . . . . . "Chaste and True."
O-Ryō . . . . . . "The Generous," — magnanimous.
O-Sada . . . . . . "The Chaste."
O-Sei . . . . . . "Truth."
O-Shin . . . . . . "Faith," — in the sense of fidelity, trust.
O-Shizu . . . . . "The Tranquil," — calm-souled.
O-Setsu . . . . . "Fidelity," — wifely virtue.
O-Tamé . . . . . "For-the-sake-of," — a name suggesting unselfishness.
O-Tei . . . . . . "The Docile," — in the meaning of virtuous obedience.
O-Toku . . . . . "Virtue."
O-Tomo . . . . . "The Friend," — especially in the meaning of mate, companion.
O-Tsuné . . . . . "Constancy."
O-Yasu . . . . . "The Amiable," — gentle.
O-Yoshi . . . . . "The Good."
O-Yoshi . . . . . "The Respectful."

The next list will appear at first sight more heterogeneous than it really is. It contains a larger variety of appellations than the previous list; but nearly all of the *yobina* refer to some good quality which the parents trust that the child will display, or to some future happiness which they hope that she will deserve. To the latter category belong such names of felicitation as *Miyo* and *Masayo*.

MISCELLANEOUS NAMES EXPRESSING PERSONAL QUALITIES, OR PARENTAL HOPES

O-Atsu	"The Generous," — liberal.
O-Chika	"Closely Dear."
O-Chika	"Thousand Rejoicings."
O-Chō	"The Long," — probably in reference to life.
O-Dai	"Great."
O-Den	"Transmission," — bequest from ancestors, tradition.
O-Ê	"Fortunate."
O-Ei	"Prosperity."
O-En	"Charm."
O-En	"Prolongation," — of life.
O-Etsu	"Surpassing."
O-Etsu	"The Playful," — merry, joyous.
O-Fuku	"Good Luck."
O-Gen	"Source," — spring, fountain.
O-Haya	"The Quick," — light, nimble.
O-Hidé	"Superior."
Hidéyo	"Superior Generations."
O-Hiro	"The Broad."
O-Hisa	"The Long." (?)
Isamu	"The Vigorous," — spirited, robust.
O-Jin	"Superexcellent."
Kaméyo	"Generations-of-the-Tortoise."
O-Kané[1]	"The Doubly-Accomplished."

[1] From the strange verb *kaneru*, signifying, to do two things at the same time.

Kaoru	"The Fragrant."
O-Kata	"Worthy Person."
O-Katsu	"The Victorious."
O-Kei	"Delight."
O-Kei	"The Respectful."
O-Ken	"The Humble."
O-Kichi	"The Fortunate."
O-Kimi	"The Sovereign," — peerless.
O-Kiwa	"The Distinguished."
O-Kiyo *Kiyoshi*	"The Clear," — in the sense of bright, beautiful.
O-Kuru	"She-who-Comes" (?).[1]
O-Maru	"The Round," — plump.
O-Masa	"The Genteel."
Masayo	"Generations-of-the-Just."
O-Masu	"Increase."
O-Mië	"Triple Branch."
O-Miki	"Stem."
O-Mio	"Triple Cord."
O-Mitsu	"Abundance."
O-Miwa	"The Far-seeing."
O-Miwa	"Three Spokes" (?).[2]
O-Miyo	"Beautiful Generations."
Miyuki[3]	"Deep Snow."
O-Moto	"Origin."

---

[1] One is reminded of, "O whistle, and I'll come to you, my lad" — but no Japanese female name could have the implied signification. More probably the reference is to household obedience.

[2] Such is the meaning of the characters. I cannot understand the name. A Buddhist explanation suggests itself; but there are few, if any, Buddhist *yobina*.

[3] This beautiful name refers to the silence and calm following a heavy snowfall. But, even for the Japanese, it is an æsthetic name also — suggesting both tranquillity and beauty.

O-Naka	"Friendship."
O-Rai	"Trust."
O-Raku[1]	"Pleasure."
O-Sachi	"Bliss."
O-Sai	"The Talented."
Sakaë	"Prosperity."
O-Saku	"The Blooming."
O-Sei	"The Refined," — in the sense of "clear."
O-Sei	"Force."
O-Sen	"Sennin," — wood-fairy.
O-Shigé	"Exuberant."
O-Shimé	"The Total," — *summum bonum.*
O-Shin	"The Fresh."
O-Shin	"Truth."
O-Shina	"Goods," — possessions.
Shirushi	"Proof," — evidence.
O-Shizu	"The Humble."
O-Shō	"Truth."
O-Shun	"Excellence."
O-Suki	"The Beloved," — *Aimée.*
O-Suké	"The Helper."
O-Sumi	"The Refined," — in the sense of "sifted."
O-Suté	"The Forsaken," — foundling.[2]

[1] The name seems curious, in view of the common proverb, *Raku wa ku no tané,* — "Pleasure is the seed of pain."

[2] Not necessarily a real foundling. Sometimes the name may be explained by a curious old custom. In a certain family several children in succession die shortly after birth. It is decided, according to traditional usage, that the next child born must be exposed. A girl is the next child born; — she is carried by a servant to some lonely place in the fields, or elsewhere, and left there. Then a peasant, or other person, hired for the occasion (it is necessary that he should be of no kin to the

O-Taë	. . . . .	"The Exquisite."
O-Taka	. . . .	"The Honorable."
O-Taka	. . . .	"The Tall."
Takara	. . . .	"Treasure," — precious object.
O-Tama	. . . .	"Jewel."
Tamaë	. . . . .	"Jewel-branch."
Tokiwa [1]	. . . .	"Eternally Constant."
O-Tomi	. . . .	"Riches."
O-Toshi	. . . .	"The Deft," — skilful.
O-Tsuma	. . . .	"The Wife."
O-Yori	. . . . .	"The Trustworthy."
O-Waka	. . . .	"The Young."

Place-names, or geographical names, are common; but they are particularly difficult to explain. A child may be called after a place because born there, or because the parental home was there, or because of beliefs belonging to the old Chinese philosophy regarding direction and position, or because of traditional

---

family), promptly appears, pretends to find the babe, and carries it back to the parental home. "See this pretty foundling," he says to the father of the girl, — "will you not take care of it?" The child is received, and named "Suté," the foundling. By this innocent artifice, it was formerly (and perhaps in some places is still) supposed that those unseen influences, which had caused the death of the other children, might be thwarted.

[1] Lit., "Everlasting-Rock," — but the ethical meaning is "Constancy-everlasting-as-the-Rocks." "Tokiwa" is a name famous both in history and tradition; for it was the name of the mother of Yoshitsuné. Her touching story, — and especially the episode of her flight through the deep snow with her boys, — has been a source of inspiration to generations of artists.

custom, or because of ideas connected with the religion of Shintō.

PLACE-NAMES

O-Fuji	[Mount] "Fuji."
O-Hama	"Coast."
O-Ichi	"Market," — fair.
O-Iyo	"Iyo," — province of Iyo, in Shikoku.
O-Kawa (rare) . .	"River."
O-Kishi	"Beach," — shore.
O-Kita	"North."
O-Kiwa	"Border."
O-Kuni	"Province."
O-Kyō	"Capital," — metropolis, — Kyōto.
O-Machi	"Town."
Matsuë	"Matsuë," — chief city of Izumo.
O-Mina [1]	"South."
O-Miné	"Peak."
O-Miya	"Temple" [*Shintō*].[2]
O-Mon [3]	"Gate."
O-Mura	"Village."
O-Nami [4]	"Wave."
Naniwa	"Naniwa," — ancient name of Ōsaka.
O-Nishi	"West."

[1] Abbreviation of *Minami*.

[2] I must confess that in classing this name as a place-name, I am only making a guess. It seems to me that the name probably refers to the *ichi no miya*, or chief Shintō temple of some province.

[3] I fancy that this name, like that of O-Séki, must have originated in the custom of naming children after the place, or neighborhood, where the family lived. But here again, I am guessing.

[4] This classification also is a guess. I could learn nothing about the name, except the curious fact that it is said to be unlucky.

O-Rin	"Park."
O-Saki	"Cape."
O-Sato	"Native Place," — village, — also, home.
O-Sawa	"Marsh."
O-Seki	"Toll-Gate," — barrier.
Shigéki	"Thickwood," — forest.
O-Shima	"Island."
O-Sono	"Flower-garden."
O-Taki	"Cataract," — or Waterfall.
O-Tani	"Valley."
O-Tsuka	"Milestone."
O-Yama	"Mountain."

The next list is a curious medley, so far as regards the quality of the *yobina* comprised in it. Some are really æsthetic and pleasing; others industrial only; while a few might be taken for nicknames of the most disagreeable kind.

NAMES OF OBJECTS AND OF OCCUPATIONS ESPECIALLY PERTAINING TO WOMEN

Ayako or *O-Aya*[1] . . .	"Damask-pattern."
O-Fumi	"Woman's Letter."
O-Fusa	"Tassel."
O-Ito	"Thread."
O-Kama[2]	"Rice-Sickle."

[1] *Aya-Nishiki*, — the famous figured damask brocade of Kyōto, — is probably referred to.

[2] *O-Kama* (Sickle) is a familiar peasant-name. *O-Kama* (caldron, or iron cooking-pot), and several other ugly names in this list are ser-

O-Kama	. . . .	"Caldron."
Kazashi	. . . .	"Hair-pin."
O-Kinu	. . . .	"Cloth-of-Silk."
O-Koto	. . . . .	"Harp."
O-Nabé	. . . . .	"Pot,"—or cooking-vessel.
O-Nui	. . . . .	"Embroidery."
O-Shimé	. . . .	"Clasp,"—ornamental fastening.
O-Somé	. . . .	"The Dyer."
O-Taru	. . . .	"Cask,"—barrel.

The following list consists entirely of material nouns used as names. There are several *yobina* among them of which I cannot find the emblematical meaning. Generally speaking, the *yobina* which signify precious substances, such as silver and gold, are æsthetic names; and those which signify common hard substances, such as stone, rock, iron, are intended to suggest firmness or strength of character. But the name "Rock" is also sometimes used as a symbol of the wish for long life, or long continuance of the family line. The curious name *Suna* has nothing, however, to do with individual "grit": it is half-moral and half-æsthetic. Fine sand—especially colored sand—is much prized in this fairy-land

vants' names. Servants in old time not only trained their children to become servants, but gave them particular names referring to their future labors.

of landscape-gardening, where it is used to cover spaces that must always be kept spotless and beautiful, and never trodden,—except by the gardener.

MATERIAL NOUNS USED AS NAMES

O-Gin	"Silver."
O-Ishi	"Stone."
O-Iwa	"Rock."
O-Kané	"Bronze."
O-Kazé[1]	"Air,"—perhaps Wind.
O-Kin	"Gold."
O-Ruri[2] / *Ruriko*	"Emerald,"—emeraldine?
O-Ryū	"Fine Metal."
O-Sato	"Sugar."
O-Seki	"Stone."
O-Shiwo	"Salt."
O-Suna	"Sand."
O-Suzu	"Tin."
O-Tané	"Seed."
O-Tetsu	"Iron."

The following five *yobina* are æsthetic names,—although literally signifying things belonging to intellectual work. Four of them, at least,

[1] I cannot find any explanation of this curious name.

[2] The Japanese name does not give the same quality of æsthetic sensation as the name Esmeralda. The *ruri* is not usually green, but blue; and the term "ruri-iro" (emerald color) commonly signifies a dark violet.

refer to calligraphy,—the matchless calligraphy of the Far East,—rather than to anything that we should call "*literary* beauty."

LITERARY NAMES

O-Bun	"Composition."
O-Fudé	"Writing-Brush."
O-Fumi	"Letter."
O-Kaku	"Writing."
O-Uta	"Poem."

Names relating to number are very common, but also very interesting. They may be loosely divided into two sub-classes,—names indicating the order or the time of birth, and names of felicitation. Such *yobina* as *Ichi*, *San*, *Roku*, *Hachi* usually refer to the order of birth; but sometimes they record the date of birth. For example, I know a person called O-Roku, who received this name, not because she was the sixth child born in the family, but because she entered this world upon the sixth day of the sixth month of the sixth Meiji. It will be observed that the numbers Two, Five, and Nine are not represented in the list: the mere idea of such names as *O-Ni*, *O-Go*, or *O-Ku* seems to a Japanese absurd. I do not know exactly why,—unless it be that they suggest unpleasant puns. The place of *O-Ni* is well supplied, however, by the name *O-Tsugi* ("Next"), which will be found in a subsequent list. Names signifying numbers ranging from eighty to a thousand, and upward, are names of felicitation. They express the wish that the bearer may live to a prodigious age, or that her posterity may flourish through the centuries.

NUMERALS AND WORDS RELATING TO NUMBER

O-Ichi	"One."
O-San	"Three."
O-Mitsu	"Three."
O-Yotsu	"Four."
O-Roku	"Six."
O-Shichi	"Seven."
O-Hachi	"Eight."
O-Jū	"Ten."
O-Iso	"Fifty." [1]
O-Yaso	"Eighty."
O-Hyaku	"Hundred." [2]
O-Yao	"Eight Hundred."
O-Sen	"Thousand."
O-Michi	"Three Thousand."
O-Man	"Ten Thousand."

[1] Such a name may record the fact that the girl was a first-born child, and the father fifty years old at the time of her birth.

[2] The "O" before this trisyllable seems contrary to rule; but *Hyaku* is pronounced almost like a dissyllable.

O-Chiyo . . . . "Thousand Generations."
Yachiyo . . . . "Eight Thousand Generations."

O-Shigé . . . . "Two-fold."
O-Yaë . . . . . "Eight-fold."
O-Kazu . . . . "Great Number."
O-Mina . . . . "All."
O-Han . . . . . "Half."[1]
O-Iku . . . . . "How Many?" (?)

OTHER NAMES RELATING TO ORDER OF BIRTH

O-Hatsu . . . . "Beginning," — first-born.
O-Tsugi . . . . "Next," — the second.
O-Naka . . . . "Midmost."
O-Tomé . . . . "Stop," — cease.
O-Sué . . . . . "Last."

Some few of the next group of names are probably æsthetic. But such names are sometimes given only in reference to the time or season of birth; and the reason for any particular *yobina* of this class is difficult to decide without personal inquiry.

NAMES RELATING TO TIME AND SEASON

O-Haru . . . . "Spring."
O-Natsu . . . . "Summer."

1 "Better half?" — the reader may query. But I believe that this name originated in the old custom of taking a single character of the father's name — sometimes also a character of the mother's name — to compose the child's name with. Perhaps in this case the name of the girl's father was HANyémon, or HANbei.

O-Aki	"Autumn."
O-Fuyu	"Winter."
O-Asa	"Morning."
O-Chō	"Dawn."
O-Yoi	"Evening."
O-Sayo	"Night."
O-Ima	"Now."
O-Toki	"Time," —opportunity.
O-Toshi	"Year [of Plenty]."

Names of animals — real or mythical — form another class of *yobina*. A name of this kind generally represents the hope that the child will develop some quality or capacity symbolized by the creature after which it has been called. Names such as "Dragon," "Tiger," "Bear," etc., are intended in most cases to represent moral rather than other qualities. The moral symbolism of the *Koi* (Carp) is too well-known to require explanation here. The names *Kamé* and *Tsuru* refer to longevity. *Koma*, curious as the fact may seem, is a name of endearment.

NAMES OF BIRDS, FISHES, ANIMALS, ETC.

Chidori	"Sanderling."
O-Kamé	"Tortoise."
O-Koi	"Carp."[1]

[1] *Cyprinus carpio.*

O-Koma	"Filly," — or pony.
O-Kuma	"Bear."
O-Ryō	"Dragon."
O-Shika	"Deer."
O-Tai	"Bream."[1]
O-Taka	"Hawk."
O-Tako	"Cuttlefish." (?)
O-Tatsu	"Dragon."
O-Tora	"Tiger."
O-Tori	"Bird."
O-Tsuru	"Stork."[2]
O-Washi	"Eagle."

Even *yobina* which are the names of flowers or fruits, plants or trees, are in most cases names of moral or felicitous, rather than of æsthetic meaning. The plumflower is an emblem of feminine virtue; the chrysanthemum, of longevity; the pine, both of longevity and constancy; the bamboo, of fidelity; the cedar, of moral rectitude; the willow, of docility and gentleness, as well as of physical grace. The symbolism of the lotos and of the cherryflower are probably familiar. But such names as *Hana* ("Blossom") and *Ben* ("Petal")

[1] *Chrysophris cardinalis.*

[2] Sometimes this name is shortened into *O-Tsu*. In Tōkyō at the present time it is the custom to drop the honorific "O" before such abbreviations, and to add to the name the suffix "chan," — as in the case of children's names. Thus a young woman may be caressingly addressed as "Tsu-chan" (for O-Tsuru), "Ya-chan" (for O-Yasu), etc.

are æsthetic in the true sense; and the Lily remains in Japan, as elsewhere, an emblem of feminine grace.

FLOWER-NAMES

Ayamé	"Iris."[1]
Azami	"Thistle-Flower."
O-Ben	"Petal."
O-Fuji	"Wistaria."[2]
O-Hana	"Blossom."
O-Kiku	"Chrysanthemum."
O-Ran	"Orchid."
O-Ren	"Lotos."
Sakurako	"Cherryblossom."
O-Umé	"Plumflower."
O-Yuri	"Lily."

NAMES OF PLANTS, FRUITS, AND TREES

O-Iné	"Rice-in-the-blade."
Kaëdé	"Maple-leaf."
O-Kaya	"Rush."[3]
O-Kaya	"Yew."[4]
O-Kuri	"Chestnut."
O-Kuwa	"Mulberry."
O-Maki	"Fir."[5]
O-Mamé	"Bean."

[1] *Iris setosa*, or *Iris sibrisia.*
[2] *Wistaria chinensis.*
[3] *Imperata arundinacea.*
[4] *Torreya nucifera.*
[5] *Podocarpus chinensis.*

O-Momo	"Peach," — the fruit.[1]
O-Nara	"Oak."
O-Ryū	"Willow."
Sanaë	"Sprouting-Rice."
O-Sané	"Fruit-seed."
O-Shino	"Slender Bamboo."
O-Sugé	"Reed."[2]
O-Sugi	"Cedar."[3]
O-Také	"Bamboo."
O-Tsuta	"Ivy."[4]
O-Yaë	"Double-Blossom."[5]
O-Yoné	"Rice-in-grain."
Wakana	"Young *Na*."[6]

Names signifying light or color seem to us the most æsthetic of all *yobina;* and they probably seem so to the Japanese. Nevertheless the relative purport even of these names cannot be divined at sight. Colors have moral and other values in the old nature-philosophy; and an appellation that to the Western mind suggests only luminosity or beauty may actually refer

---

[1] Yet this name may possibly have been written with the wrong character. There is another *yobina,* "Momo" signifying "hundred," — as in the phrase *momo yo,* "for a hundred ages."

[2] *Scirpus maritimus.*

[3] *Cryptomeria Japonica.*

[4] *Cissus Thunbergii.*

[5] A flower-name certainly; but the *yaë* here is probably an abbreviation of *yaë-zakura,* the double-flower of a particular species of cherry-tree.

[6] *Brassica chinensis.*

to moral or social distinction, — to the hope that the girl so named will become " illustrious."

NAMES SIGNIFYING BRIGHTNESS

O-Mika	" New Moon." [1]
O-Mitsu	" Light."
O-Shimo	" Frost."
O-Teru	" The Shining."
O-Tsuki	" Moon."
O-Tsuya	" The Glossy," — lustrous.
O-Tsuyu	" Dew."
O-Yuki	" Snow."

COLOR-NAMES

O-Ai	" Indigo."
O-Aka	" Red."
O-Iro	" Color."
O-Kon	" Deep Blue."
O-Kuro	" Dark," — lit., " Black."
Midori [2]	" Green."
Murasaki [2] . . .	" Purple."
O-Shiro	" White."

---

[1] *Mika* is an abbreviation of *Mikazuki*, " the moon of the third night " [of the old lunar month].

[2] *Midori* and *Murasaki*, especially the latter, should properly be classed with aristocratic *yobina* ; and both are very rare. I could find neither in the collection of aristocratic names which was made for me from the records of the Peeresses' School ; but I discovered a " Midori " in a list of middle-class names. Color-names being remarkably few among *yobina*, I thought it better in this instance to group the whole of them together, independently of class-distinctions.

The following and final group of female names contains several queer puzzles. Japanese girls are sometimes named after the family crest; and heraldry might explain one or two of these *yobina*. But why a girl should be called a ship, I am not sure of being able to guess. Perhaps some reader may be reminded of Nietzsche's "Little Brig called Angeline": —

> "Angeline — they call me so —
> Now a ship, one time a maid,
> (Ah, and evermore a maid!)
> Love the steersman, to and fro,
> Turns the wheel so finely made."

But such a fancy would not enter into a Japanese mind. I find, however, in a list of family crests, two varieties of design representing a ship, twenty representing an arrow, and two representing a bow.

NAMES DIFFICULT TO CLASSIFY OR EXPLAIN

O-Fuku[1]	"Raiment," — clothing.
O-Funé	"Ship," — or Boat.
O-Hina[2]	"Doll," — a paper doll?

---

[1] Possibly this name belongs to the same class as *O-Nui* ("Embroidery"), *O-Somé* ("The Dyer"); but I am not sure.

[2] Probably a name of caress. The word *hina* is applied especially to the little paper dolls made by hand for amusement, — representing

O-Kono	. . . .	"This."
O-Nao	. . . . .	"Still More."
O-Nari	. . . . .	"Thunder-peal."
O-Nibo	. . . . .	"Palanquin," (?).
O-Rai	. . . . .	"Thunder."
O-Rui	. . . . .	"Sort," — kind, species.
O-Suzu[1]	. . . .	"Little Bell."
Suzuë	. . . . .	"Branch-of-Little-Bells."
O-Tada	. . . .	"The Only."
Tamaki	. . . .	"Armlet," — bracelet.
O-Tami	. . . .	"Folk," — common people.
O-Toshi	. . . .	"Arrowhead," — or barb.
O-Tsui	. . . . .	"Pair," — match.
O-Tsuna	. . . .	"Rope," — bond.
O-Yumi	. . . .	"Bow," — weapon.

Before passing on to the subject of aristocratic names, I must mention an old rule for Japanese names, — a curious rule that might help to account for sundry puzzles in the preceding lists. This rule formerly applied to all personal names, — masculine or feminine. It cannot be fully explained in the present paper; for a satisfactory

young ladies with elaborate coiffure; and it is also given to the old-fashioned dolls representing courtly personages in full ceremonial costume. The true doll — doll-baby — is called *ningyō*.

[1] Perhaps this name is given because of the sweet sound of the *suzu*, — a tiny metal ball, with a little stone or other hard object inside, to make the ringing. — It is a pretty Japanese custom to put one of these little *suzu* in the silk charm-bag (*mamori-bukero*) which is attached to a child's girdle. The *suzu* rings with every motion that the child makes, — somewhat like one of those tiny bells which we attach to the neck of a pet kitten.

PHONETIC RELATION OF THE FIVE ELEMENTAL-NATURES TO THE JAPANESE SYLLABARY

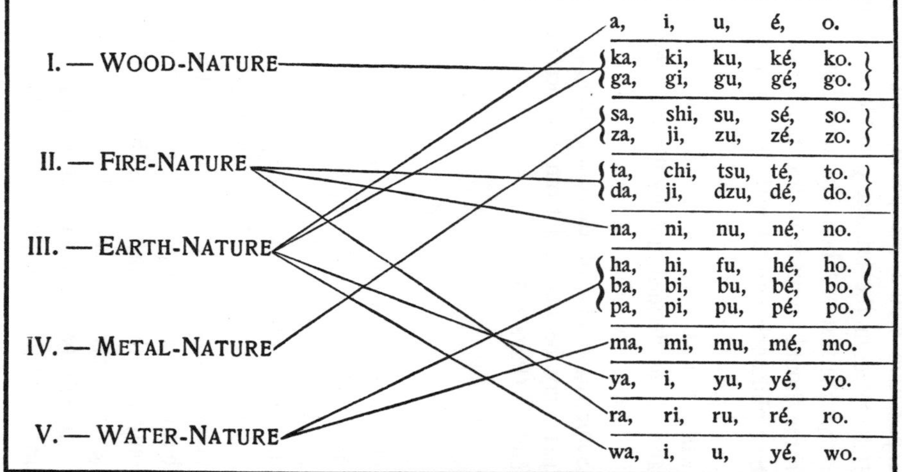

explanation would occupy at least fifty pages. But, stated in the briefest possible way, the rule is that the first or "head-character" of a personal name should be made to "accord" (in the Chinese philosophic sense) with the supposed *Sei*, or astrologically-determined nature, of the person to whom the name is given; — the required accordance being decided, not by the meaning, but by the sound of the Chinese written character. Some vague idea of the difficulties of the subject may be obtained from the accompanying table. (Page 143.)

III

FOR examples of contemporary aristocratic names I consulted the reports of the *Kwazoku-Jogakkō* (Peeresses' School), published between the nineteenth and twenty-seventh years of Meiji (1886–1895). The Kwazoku-Jogakkō admits other students besides daughters of the nobility; but for present purposes the names of the latter only — to the number of one hundred and forty-seven — have been selected.

It will be observed that names of three or more syllables are rare among these, and also that the modern aristocratic *yobina* of two syllables, as pronounced and explained, differ little from ordinary *yobina*. But as written in Chinese they differ greatly from other female names, being in most cases represented by characters of a complex and unfamiliar kind. The use of these more elaborate characters chiefly accounts for the relatively large number of homonyms to be found in the following list: —

PERSONAL NAMES OF LADY STUDENTS OF THE KWAZOKU JOGAKKŌ

Aki-ko	"Autumn."
Aki-ko	"The Clear-Minded."
Aki-ko	"Dawn."
Asa-ko	"Fair Morning."
Aya-ko	"Silk Damask."
Chiharu-ko . . .	"A Thousand Springs."
Chika-ko	"Near," — close.
Chitsuru-ko . . .	"A Thousand Storks."
Chiyo-ko	"A Thousand Generations."
Ei-ko	"Bell-Chime."
Etsu-ko	"Delight."
Fuji-ko	"Wistaria."
Fuku-ko	"Good-Fortune."
Fumi-ko	"A Woman's Letter."
Fuyō-ko	"Lotos-flower."
Fuyu-ko	"Winter."
Hana-ko	"Flower."

Hana-ko . . . . "Fair-Blooming."
Haru-ko . . . . "The Tranquil."
Haru-ko . . . . "Spring,"—the season of flowers.
Haru-ko . . . . "The Far-Removed,"—in the sense, perhaps, of superlative.
Hatsu-ko . . . . "The First-born."
Hidé-ko . . . . "Excelling."
Hidé-ko . . . . "Surpassing."
Hiro-ko . . . . . "Magnanimous,"—literally, "broad," "large,"—in the sense of beneficence.
Hiro-ko . . . . . "Wide-Spreading,"—with reference to family prosperity.
Hisa-ko . . . . . "Long-lasting."
Hisa-ko . . . . . "Continuing."
Hoshi-ko . . . . "Star."
Iku-ko . . . . . "The Quick,"—in the sense of living.
Ima-ko . . . . . "Now."
Iho-ko . . . . . "Five Hundred,"—probably a name of felicitation.
Ito-ko . . . . . "Sewing-Thread."
Kamé-ko . . . . "Tortoise."
Kané-ko . . . . "Going around" (?).[1]
Kané-ko . . . . "Bell,"—the character indicates a large suspended bell.
Kata-ko . . . . "Condition"?
Kazu-ko . . . . "First."
Kazu-ko . . . . "Number,"—a great number.
Kazu-ko . . . . "The Obedient."
Kiyo-ko . . . . "The Pure."

[1] It is possible that this name was made simply by taking one character of the father's name. The girl's name otherwise conveys no intelligible meaning.

Kō[1]	"Filial Piety."
Kō-ko	"Stork."
Koto	"Harp."
Kuni-ko	"Province."
Kuni	"Country," — in the largest sense.
Kyō-ko	"Capital," — metropolis.
Machi	"Ten-Thousand Thousand."
Makoto	"True-Heart."
Masa-ko	"The Trustworthy," — sure.
Masa-ko	"The Upright."
Masu-ko	"Increase."
Mata-ko	"Completely," — wholly.
Matsu-ko	"Pine-tree."
Michi-ko	"Three Thousand."
Miné	"Peak."
Miné-ko	"Mountain-Range."
Mitsu-ko	"Light," — radiance.
Miyo-ko	"Beautiful Generations."
Moto-ko	"Origin," — source.
Naga-ko	"Long," — probably in reference to time.
Naga-ko	"Long Life."
Nami-ko	"Wave."
Nao-ko	"Correct," — upright.
Nyo-ko[2]	"Gem-Treasure."

[1] The suffix "*ko*" is sometimes dropped for reasons of euphony, and sometimes for reasons of good taste — difficult to explain to readers unfamiliar with the Japanese language — even when the name consists of only one syllable or of two syllables.

[2] This name is borrowed from the name of the sacred gem *Nyoibōju*, which figures both in Shintō and in Buddhist legend. The divinity Jizō is usually represented holding in one hand this gem, which is said to have the power of gratifying any desire that its owner can entertain. Perhaps the *Nyoibōju* may be identified with the Gem-Treasure *Veluriya*,

Nobu-ko	"Faithful."
Nobu-ko	"Abundance,"—plenty.
Nobu-ko	"The Prolonger."
Nori-ko	"Precept,"—doctrine.
Nui	"Embroidery,"—sewing.
Oki	"Offing,"—perhaps originally a place-name.[1]
Sada-ko	"The Chaste."
Sada-ko	"The Sure,"—trustworthy.
Sakura-ko	"Cherry-Blossom."
Sakaë	"The Prosperous."
Sato-ko	"Home."
Sato-ko	"The Discriminating."
Seki-ko	"Great."
Setsu-ko	"The Chaste."
Shigé-ko	"Flourishing."
Shigé-ko	"Exuberant,"—in the sense of rich growth.
Shigé-ko	"Upgrowing."
Shigé-ko	"Fragrance."
Shiki-ko	"Prudence."
Shima-ko	"Island."
Shin-ko	"The Fresh,"—new.
Shizu-ko	"The Quiet,"—calm.
Shizuë	"Quiet River."
Sono-ko	"Garden."
Suë-ko	"Last,"—in the sense of youngest.
Suké-ko	"The Helper."

mentioned in the Sûtra of The Great King of Glory, chapter i. (See *Sacred Books of the East*, vol. xi.)

[1] A naval officer named Oki told me that his family had originally been settled in the Oki Islands ("Islands of the Offing"). This interesting coincidence suggested to me that the above *yobina* might have had the same origin.

Sumi-ko . . . . "The Clear," — spotless, refined.
Sumi-ko . . . . "The Veritable," — real.
Sumië-ko . . . . "Clear River."
Suzu-ko . . . . "Tin."
Suzu-ko . . . . "Little Bell."
Suzunë . . . . . "Sound of Little Bell."
Taka-ko . . . . "High," — lofty, superior.
Taka-ko . . . . "Filial Piety."
Taka-ko . . . . "Precious."
Také-ko . . . . "Bamboo."
Taki-ko . . . . "Waterfall."
Tama-ko . . . . "Gem," — jewel.
Tama-ko . . . . "Gem," — written with a different character.
Tamé-ko . . . . "For the Sake of —"
Tami-ko . . . . "People," — folks.
Tané-ko . . . . "Successful."
Tatsu-ko . . . . "Attaining."
Tatsuru-ko[1] . . . "Many Storks."
Tatsuru-ko . . . "Ricefield Stork."
Teru-ko . . . . "Beaming," — luminous.
Tetsu-ko . . . . "Iron."
Toki-ko . . . . . "Time."
Tomé-ko . . . . "Cessation."
Tomi-ko . . . . "Riches."
Tomo . . . . . "Intelligence."
Tomo . . . . . "Knowledge."
Tomo-ko . . . . "Friendship."
Toshi-ko . . . . "The Quickly-Perceiving."
Toyo-ko . . . . "Fruitful."
Tsuné . . . . . "Constancy."
Tsuné-ko . . . . "Ordinary," — usual, common.

---

[1] So written, but probably pronounced as two syllables only.

Tsuné-ko . . . . "Ordinary,"—written with a different character.
Tsuné-ko . . . . "Faithful,"—in the sense of wifely fidelity.
Tsuru-ko . . . . "Stork."
Tsuya-ko . . . . "The Lustrous,"—shining, glossy.
Umé . . . . . . "Female Hare."
Umé-ko . . . . "Plum-Blossom."
Yachi-ko . . . . "Eight Thousand."
Yaso-ko . . . . "Eighty."
Yasoshi-ko . . . "Eighty-four."
Yasu-ko . . . . "The Maintainer,"—supporter.
Yasu-ko . . . . "The Respectful."
Yasu-ko . . . . "The Tranquil-Minded."
Yoné-ko . . . . "Rice."
Yori-ko . . . . "The Trustful."
Yoshi . . . . . "Eminent,"—celebrated.
Yoshi-ko . . . . "Fragrance."
Yoshi-ko . . . . "The Good,"—or Gentle.
Yoshi-ko . . . . "The Lovable."
Yoshi-ko . . . . "The Lady-like,"—gentle in the sense of refined.
Yoshi-ko . . . . "The Joyful."
Yoshi-ko . . . . "Congratulation."
Yoshi-ko . . . . "The Happy."
Yoshi-ko . . . . "Bright and Clear."
Yuki-ko . . . . "The Lucky."
Yuki-ko . . . . "Snow."
Yuku-ko . . . . "Going."
Yutaka . . . . . "Plenty,"—affluence, superabundance.

IV

In the first part of this paper I suggested that the custom of giving very poetical names to *geisha* and to *jorō* might partly account for the unpopularity of purely æsthetic *yobina*. And in the hope of correcting certain foreign misapprehensions, I shall now venture a few remarks about the names of *geisha*.

Geisha-names, — like other classes of names, — although full of curious interest, and often in themselves really beautiful, have become hopelessly vulgarized by association with a calling the reverse of respectable. Strictly speaking, they have nothing to do with the subject of the present study, — inasmuch as they are not real personal names, but professional appellations only, — not *yobina*, but *geimyō*.

A large proportion of such names can be distinguished by certain prefixes or suffixes attached to them. They can be known, for example, —

(1) By the prefix *Waka*, signifying "Young"; — as in the names *Wakagusa*, "Young Grass"; *Wakazuru*, "Young Stork"; *Wakamurasaki*, "Young Purple"; *Wakakoma*, "Young Filly".

(2) By the prefix *Ko*, signifying " Little " ;— as in the names, *Ko-en*, " Little Charm " ; *Ko-hana*, " Little Flower "; *Kozakura*, " Little Cherry-Tree ".

(3) By the suffix *Ryō*, signifying " Dragon " (the Ascending Dragon being especially a symbol of success) ;— as *Tama-Ryō*, " Jewel-Dragon "; *Hana-Ryō*, " Flower-Dragon " ; *Kin-Ryō*, " Golden-Dragon ".

(4) By the suffix *ji*, signifying " to serve ", " to administer " ;— as in the names *Uta-ji*, *Shinné-ji*, *Katsu-ji*.

(5) By the suffix *suké*, signifying " help " ;— as in the names *Tama-suké*, *Koma-suké*.

(6) By the suffix *kichi*, signifying " luck ", " fortune " ; — as *Uta-kichi*, " Song-Luck "; *Tama-kichi*, " Jewel-Fortune ".

(7) By the suffix *giku* (i. e., *kiku*), signifying " chrysanthemum " ;— as *Mitsu-giku*, " Three Chrysanthemums " ; *Hina-giku*, " Doll-Chrysanthemum "; *Ko-giku*, " Little Chrysanthemum ".

(8) By the suffix *tsuru*, signifying " stork " (emblem of longevity) ; — as *Koma-tsuru*, " Filly-Stork " ; *Ko-tsuru*, " Little Stork " ; *Ito-zuru*, " Thread-Stork ".

These forms will serve for illustration; but there are others. *Geimyō* are written, as a general rule, with only two Chinese characters, and are pronounced as three or as four syllables. *Geimyō* of five syllables are occasionally to be met with; *geimyō* of only two syllables are rare — at least among names of dancing girls. And these professional appellations have seldom any moral meaning: they signify things relating to longevity, wealth, pleasure, youth, or luck, — perhaps especially to luck.

Of late years it became a fashion among certain classes of *geisha* in the capital to assume real names with the genteel suffix *Ko*, and even aristocratic *yobina*. In 1889 some of the Tōkyō newspapers demanded legislative measures to check the practice. This incident would seem to afford proof of public feeling upon the subject.

Old Japanese Songs

Old Japanese Songs

THIS New Year's morning I find upon my table two most welcome gifts from a young poet of my literary class. One is a roll of cloth for a new kimono, — cloth such as my Western reader never saw. The brown warp is cotton thread; but the woof is soft white paper string, irregularly speckled with black. When closely examined, the black specklings prove to be Chinese and Japanese characters; — for the paper woof is made out of manuscript, — manuscript of poems, — which has been deftly twisted into fine cord, with the written surface outwards. The general effect of the white, black, and brown in the texture is a warm mouse-grey. In many Izumo homes a similar kind of cloth is manufactured for family use; but this piece was woven especially for me by the mother of my pupil. It will make a most comfortable winter-robe; and when wearing it, I shall be literally clothed with poetry,—even as a divinity might be clothed with the sun.

The other gift is poetry also, but poetry in the original state: a wonderful manuscript collection of Japanese songs gathered from unfamiliar sources, and particularly interesting from the fact that nearly all of them are furnished with refrains. There are hundreds of compositions, old and new,—including several extraordinary ballads, many dancing-songs, and a surprising variety of love-songs. Neither in sentiment nor in construction do any of these resemble the Japanese poetry of which I have already, in previous books, offered specimens in translation. The forms are, in most cases, curiously irregular; but their irregularity is not without a strange charm of its own.

I am going to offer examples of these compositions,—partly because of their unfamiliar emotional quality, and partly because I think that something can be learned from their strange art of construction. The older songs—selected from the antique drama—seem to me particularly worthy of notice. The thought or feeling and its utterance are supremely simple; yet by primitive devices of reiteration and of pause, very remarkable results have been obtained. What strikes me especially noteworthy in the following specimen is the way that the phrase, begun with the third line of the first stanza, and interrupted by a kind of burthen, is repeated and finished in the next stanza. Perhaps the suspension will recall to Western readers the effect of some English ballads with double refrains, or of such quaint forms of French song as the famous —

Au jardin de mon père —
    Vole, mon cœur, vole!
Il y a un pommier doux,
    Tout doux!

But in the Japanese song the reiteration of the broken phrase produces a slow dreamy effect as unlike the effect of the French composition as the movements of a Japanese dance are unlike those of any Western round: —

KANO YUKU WA

(Probably from the eleventh century)

Kano yuku wa,
Kari ka ? — kugui ka ?
Kari naraba, —

(Ref.) *Harėya tōtō !*
Harėya tōtō !

Kari nara
Nanori zo sémashi ; —
Nao kugui nari-ya ! —

(Ref.) *Tōtō !*

That which yonder flies, —
Wild goose is it ? — swan is it ?
Wild goose if it be, —
Haréya tōtō !
Haréya tōtō !

Wild goose if it be,
Its name I soon shall say :
Wild swan if it be, — better still !
Tōtō !

There are many old lyrics in the above form. Here is another song, of different construction, also from the old drama : there is no refrain, but there is the same peculiar suspension of phrase; and the effect of the quadruple repetition is emotionally impressive:—

Isora ga saki ni
Tai tsuru ama mo,
Tai tsuru ama mo,—

Wagimoko ga tamé to,
Tai tsuru ama mo,
Tai tsuru ama mo!

Off the Cape of Isora,
Even the fisherman catching *tai*,[1]
Even the fisherman catching *tai*,—

[Works] for the sake of the woman beloved,—
Even the fisherman catching *tai*,
Even the fisherman catching *tai!*

But a still more remarkable effect is obtained in the following ancient song by the extraordinary reiteration of an uncompleted phrase, and by a double suspension. I can imagine nothing more purely natural: indeed the realism of these simple utterances has almost the quality of pathos:—

[1] *Chrysophris cardinalis*, a kind of sea-bream,—generally esteemed the best of Japanese fishes.

AGÉMAKI

(*Old lyrical drama — date uncertain*)

Agémaki[1] wo
Waséda ni yarité ya!
So omou to,
So omou to,
So omou to,
So omou to,
So omou to, —

So omou to,
Nani-mo sezushité, —
Harubi sura,
Harubi sura,
Harubi sura,
Harubi sura,
Harubi sura!

My darling boy! —
Oh! they have sent him to the ricefields!
When I think about him, —
When I think,
When I think,

---

[1] It was formerly the custom to shave the heads of boys, leaving only a tuft or lock of hair on either temple. Such a lock was called *agémaki*, a word also meaning "tassel"; and eventually the term came to signify a boy or lad. In these songs it is used as a term of endearment, — much as an English girl might speak of her sweetheart as "my dear lad," or "my darling boy."

When I think,
When I think, —

When I think about him!
I — doing nothing at all, —
Even on this spring-day,
Even this spring-day,
Even this spring-day,
Even this spring-day,
Even on this spring-day! —

Other forms of repetition and of refrain are furnished in the two following lyrics: —

BINDATARA

(Supposed to have been composed as early as the twelfth century)

Bindatara wo
Ayugaséba koso,
Ayugaséba koso,
Aikyō zuitaré!
Yaréko tōtō,
Yaréko tōtō!

With loosened hair, —
Only because of having tossed it,
Only because of having shaken it, —
Oh, sweet she is!
Yaréko tōtō!
Yaréko tōtō!

SAMA WA TENNIN

(*Probably from the sixteenth century*)

Sama wa tennin!
Soré-soré,
Tontorori!

Otomé no sugata
Kumo no kayoiji
Chirato mita!
Tontorori!

Otomé no sugata
Kumo no kayoiji
Chirato mita!
Tontorori!

My beloved an angel is![1]
Soré-soré!
Tontorori!

The maiden's form,
In the passing of clouds,
In a glimpse I saw!
Tontorori!

The maiden's form,
In the passage of clouds,
In a glimpse I saw!
Tontorori!

---

[1] Lit., "a Tennin";—that is to say, an inhabitant of the Buddhist heaven. The Tennin are usually represented as beautiful maidens.

My next selection is from a love-song of uncertain date, belonging to the Kamakura period (1186–1332). This fragment is chiefly remarkable for its Buddhist allusions, and for its very regular form of stanza: —

Makoto yara,
Kashima no minato ni
Miroku no mifuné ga
Tsuité gozarimōsu.

Yono!
Sā iyoë, iyoë!
Sā iyoë, iyoë!

Hobashira wa,
Kogané no hobashira;
Ho niwa Hokkékyō no
Go no man-makimono.

Sā iyoë, iyoë!
Sā iyoë, iyoë!

. . . . .

I know not if 't is true
That to the port of Kashima
The august ship of Miroku[1] has come!

Yono!
Sā iyoë, iyoë!
Sā iyoë, iyoë!

---

[1] Miroku Bosatsu (Maitrêya Bodhisattva) is the next great Buddha to come.

As for the mast,
It is a mast of gold; —
The sail is the fifth august roll
Of the Hokkékyō ! [1]

Sā iyoë, iyoë !
Sā iyoë, iyoë

. . . .

Otherwise interesting, with its queer refrain, is another song called "Agémaki," — belonging to one of the curious class of lyrical dramas known as *Saibara.* This may be found fault with as somewhat "free"; but I cannot think it more open to objection than some of our much-admired Elizabethan songs which were probably produced at about the same time: —

AGÉMAKI

(*Probably from the sixteenth century*)

Agémaki ya !
Tonton !
Hiro bakari ya —
Tonton !

---

[1] Japanese popular name for the Chinese version of the Saddhârma Pundarîka Sûtra. — Many of the old Buddhist scriptures were written upon long scrolls, called *makimono,* — a name also given to pictures printed upon long rolls of silk or paper.

Sakarité netarédomo,
Marobi-ainikéri,—
Tonton!
Kayori-ainikéri,
Tonton!

Oh! my darling boy!
Tonton!
Though a fathom[1] apart,
Tonton!
Sleeping separated,
By rolling we came together!
Tonton!
By slow approaches we came together,
Tonton!

My next group of selections consists of "local songs"—by which term the collector means songs peculiar to particular districts or provinces. They are old—though less old than the compositions previously cited;—and their interest is chiefly emotional. But several, it will be observed, have curious refrains. Songs of this sort are sung especially at the village-dances—*Bon-odori* and *Hōnen-odori*:—

---

[1] Lit., "*hiro*." The *hiro* is a measure of about five feet English, and is used to measure breadth as well as depth.

LOVE-SONG

(*Province of Echigo*)

Hana ka?—chōchō ka?
Chōchō ka?—hana ka?
Don-don!
Kité wa chira-chira mayowaséru,
Kité wa chira-chira mayowaséru!
Taichokanė!
Sōkanė don-don!

Flower is it?—butterfly is it?
Butterfly or flower?
Don-don!
When you come thus flickering, I am deluded!—
When you come thus twinkling, I am bewitched!
Taichokané!
Sōkané don-don!

LOVE-SONG

(*Province of Kii,—village of Ogawa*)

Koë wa surédomo
Sugata wa miénu—
Fuka-no no kirigirisu!

Though I hear the voice [*of the beloved*], the form I cannot see—a *kirigirisu*[1] in the high grass.

[1] The *kirigirisu* is a kind of grasshopper with a very musical note. It is very difficult to see it, even when it is singing close by, for its color is exactly the color of the grass. The song alludes to the happy peasant custom of singing while at work in the fields.

LOVE-SONG

(*Province of Mutsu,—district of Sugaru*)

Washi no kokoro to
Oki kuru funé wa,
Raku ni misétémo,
Ku ga taënu.

My heart and a ship in the offing—either seems to move with ease; yet in both there is trouble enough.

LOVE-SONG

(*Province of Suwō,—village of Iséki*)

Namida koboshité
Shinku wo kataru,
Kawairashi-sa ga
Mashimasuru!

As she tells me all the pain of her toil, shedding tears,—ever her sweetness seems to increase.

LOVE-SONG

(*Province of Suruga, village of Gotemba*)

Hana ya, yoku kiké!
Shō aru naraba,
Hito ga fusagu ni
Nazé hiraku?

O flower, hear me well if thou hast a soul! When any one sorrows as I am sorrowing, why dost thou bloom?

OLD TŌKYŌ SONG

Iya-na o-kata no
Shinsetsu yori ka
Suita o-kata no
Muri ga yoi.

Better than the kindness of the disliked is the violence of the beloved.

LOVE-SONG

(*Province of Iwami*)

Kawairashi-sa ya!
Hotaru no mushi wa
Shinobu nawaté ni
Hi wo tomosu.

Ah, the darling! . . . Ever as I steal along the ricefield-path [*to meet my lover*], the firefly kindles a light to show me the way.

COMIC SONG

(*Province of Shinano*)

Ano yama kagé dé
Hikaru wa nanja?—
Tsuki ka, hoshi ka, hotaru no mushi ka?
Tsuki démo naiga;
Hoshi démo naiga;—
Shūto no o-uba no mé ga hikaru,—
(Chorus) *Mé ga hikaru!*

In the shadow of the mountain
What is it that shines so?
Moon is it, or star?—or is it the firefly-insect?
Neither is it moon,
Nor yet star;—
It is the old woman's Eye;—it is the Eye of my mother-in-law that shines,—
(Chorus) *It is her Eye that shines!*

KAËRI-ODORI[1]

(*Province of Sanuki*)

Oh! the cruelty, the cruelty of my mother-in-law!—
(Chorus) *Oh! the cruelty!*

Even tells me to paint a picture on running water!
If ever I paint a picture on running water,
You will count the stars in the night-sky!
Count the stars in the night-sky!

—Come! let us dance the Dance of the Honorable Garden!—
Chan-chan!
Cha-cha!
Yoitomosė,
Yoitomosė!

[1] I am not sure of the real meaning of the name *Kaëri-Odori* (lit. "turn-dance" or "return-dance").

Who cuts the bamboo at the back of the house?—

(Chorus) *Who cuts the bamboo?—*

My sweet lord's own bamboo, the first he planted,—

The first he planted?

—*Come! let us dance the Dance of the Honorable Garden!—*

Chan-chan!
Cha-cha!
Yoitomosé,
Yoitomosé!

Oh! the cruelty, the cruelty of my mother-in-law!—

Oh! the cruelty!

Tells me to cut and make a hakama[1] out of rock!
If ever I cut and sew a hakama of rock,
Then you will learn to twist the fine sand into thread,—

Twist it into thread.

—*Come! let us dance the Dance of the Honorable Garden!—*

[1] A divided skirt of a peculiar form, worn formerly by men chiefly, to-day worn by female students also.

Chan-chan!
Cha-cha!
Yoitomosé,
Yoitomosé!
Chan-chan-chan!

OTERA-ODORI (TEMPLE-DANCE)

(*Province of Iga, village called Uenomachi*)

Visiting the honorable temple, when I see the august gate,
The august gate I find to be of silver, the panels of gold.
Noble indeed is the gate of the honorable temple, —
The honorable temple!

Visiting the honorable temple, when I see the garden,
I see young pinetrees flourishing in the four directions:
On the first little branch of one the *shijūgara*[1] has made her nest, —
Has made her nest.

[1] The Manchurian great tit. It is said to bring good fortune to the owners of the garden in which it builds a nest, — providing that the nest be not disturbed and that the brood be protected.

Visiting the honorable temple, when I see the
water-tank,
I see little flowers of many colors set all about it,
Each one having a different color of its own,—
A different color.

Visiting the honorable temple, when I see the
parlor-room,
I find many kinds of little birds gathered all
together,
Each one singing a different song of its own,—
A different song.

Visiting the honorable temple, when I see the
guest-room,
There I see the priest, with a lamp beside him,
Reading behind a folding-screen—oh, how admirable it is!—
How admirable it is!

Many kinds of popular songs—and especially the class of songs sung at country-dances—are composed after a mnemonic plan. The stanzas are usually ten in number; and the first syllable of each should correspond in sound to the first syllable of the numeral placed before the verse.

Sometimes Chinese numerals are used; sometimes Japanese. But the rule is not always perfectly observed. In the following example it will be observed that the correspondence of the first two syllables in the first verse with the first two syllables of the Japanese word for one (*hitotsu*) is a correspondence of meaning only; — *ichi* being the Chinese numeral: —

SONG OF FISHERMEN

(*Province of Shimosa,— town of Chōshi*)[1]

Hitotsutosé, —
Ichiban buné é tsumi-kondé,
Kawaguchi oshikomu ō-yagoë.
Kono tai-ryō-buné!

Futatsutosé, —
Futaba no oki kara Togawa madé
Tsuzuité oshikomu ō-yagoë.
Kono tai-ryō-buné!

---

[1] Chōshi, a town of some importance, is situated at the mouth of the Tonégawa. It is celebrated for its *iwashi*-fishery. The *iwashi* is a fish about the size of the sardine, and is sought chiefly for the sake of its oil. Immense quantities of *iwashi* are taken off the coast. They are boiled to extract the oil; and the dried residue is sent inland to serve as manure.

Mitsutosé, —
Mina ichidō-ni manéki wo agé,
Kayowasé-buné no nigiyakasa
Kono tai-ryō-buné!

Yotsutosé, —
Yoru-hiru taitémo taki-amaru,
San-bai itchō no ō-iwashi!
Kono tai-ryō-buné!

Itsutsutosé, —
Itsu kité mitémo hoshika-ba ni
Akima sukima wa sarani nai.
Kono tai-ryō-buné!

Mutsutoyé, —
Mutsu kara mutsu madé kasu-wari ga
Ō-wari ko-wari dé té ni owaré.
Kono tai-ryō-buné!

Nanatsutosé, —
Natakaki Tonégawa ichi-men ni
Kasu-ya abura wo tsumi-okuru
Kono tai-ryō-buné!

Yatsutosé, —
Yatébuné no okiai wakashu ga,
Ban-shuku soroété miya-mairi.
Kono tai-ryō-buné!

Kokonotsutosé, —
Kono ura mamoru kawa-guchi no
Myōjin riyaku wo arawasuru.
Kono tai-ryo-buné!

Firstly (or "Number One"), —
The first ship, filled up with fish, squeezes her way through the river-mouth, with a great shouting.[1]
O this ship of great fishing![2]

Secondly, —
From the offing of Futaba even to the Togawa,[3] the ships, fast following, press in, with a great shouting.
O this ship of great fishing!

Thirdly, —
When, all together, we hoist our signal-flags, see how fast the cargo-boats come hurrying!
O this ship of great fishing!

Fourthly, —
Night and day though the boiling be, there is still too much to boil — oh, the heaps of *iwashi* from the three ships together!
O this ship of great fishing!

---

[1] *Ō-yagoë.* The chorus-cry or chant of sailors, pulling all together, is called *yagoë.*

[2] *Tai-ryō buné,* lit.: — "great-fishing," or "great-catching-ship." The adjective refers to the fishing, not to the ship. The real meaning of the refrain is, "this-most-successful-in-fishing of ships."

[3] Perhaps the reference is to a village at the mouth of the river Togawa, — not far from Chōshi on the Tonégawa. The two rivers are united by a canal. But the text leaves it uncertain whether river or village is meant.

Fifthly, —

Whenever you go to look at the place where the dried fish are kept,[1] never do you find any room, — not even a crevice.

O this ship of great fishing!

Sixthly,

From six to six o'clock is cleaning and washing: the great cutting and the small cutting are more than can be done.

O this ship of great fishing!

Seventhly, —

All up and down the famous river Tonégawa we send our loads of oil and fertilizer.

O this ship of great fishing!

Eighthly, —

All the young folk, drawing the *Yatai-buné*,[2] with ten thousand rejoicings, visit the shrine of the God.

O this ship of great fishing!

Ninthly, —

Augustly protecting all this coast, the Deity of the river-mouth shows to us his divine favor.

O this ship of great fishing!

---

[1] *Hoshika-ba*: lit., "the hoshika-place" or "hoshika-room." "Hoshika" is the name given to dried fish prepared for use as fertilizer.

[2] *Yatai* is the name given to the ornamental cars drawn with ropes in a religious procession. *Yatai-buné* here seems to mean either the model of a boat mounted upon such a car, or a real boat so displayed in a religious processsion. I have seen real boats mounted upon festival-cars in a religious procession at Mionoséki.

A stranger example of this mnemonic arrangement is furnished by a children's song, composed at least a hundred years ago. Little girls of Yedo used to sing it while playing ball. You can see the same ball-game being played by girls to-day, in almost any quiet street of Tōkyō. The ball is kept bounding in a nearly perpendicular line by skilful taps of the hand delivered in time to the measure of a song; and a good player should be able to sing the song through without missing a stroke. If she misses, she must yield the ball to another player.[1] There are many pretty "ball-play songs;" but this old-fashioned and long-forgotten one is a moral curiosity:—

Hitotsu to ya:—

Hito wa kō na hito to iu;
On wo shiranéba kō naraji.

Futatsu to ya:—

Fuji yori takaki chichi no on;
Tsuné-ni omouté wasuré-naji.

---

[1] This is the more common form of the game; but there are many other forms. Sometimes two girls play at once with the same ball—striking it alternately as it bounds.

Mitsu to ya : —

Mizu-umi kaetté asashi to wa,
Haha no on zo ya omou-beshi.

Yotsu to ya : —

Yoshiya mazushiku kurasu tomo,
Sugu-naru michi wo maguru-moji.

Itsutsu to ya : —

Itsumo kokoro no kawaranu wo,
Makoto no hito to omou-beshi.

Mutsu to ya : —

Munashiku tsukihi wo kurashi-naba,
Nochi no nagéki to shirinu-beshi.

Nanatsu to ya : —

Nasaki wa hito no tamé narodé,
Waga mi no tamé to omou-beshi.

Yatsu to ya : —

Yaku-nan muryō no wazawai mo
Kokoro zen nara nogaru-beshi.

Kokonotsu to ya : —

Kokoro kotoba no sugu-naraba,
Kami ya Hotoké mo mamoru-beshi.

Tō to ya:—

Tōtoi hito to naru naraba,
Kōkō mono to iwaru-beshi.

This is the first:—

[Only] a person having filial piety is [worthy to be] called a person:[1]
If one does not know the goodness of parents, one has not filial piety.

The second:—

Higher than the [mountain] Fuji is the favor of a father:
Think of it always; — never forget it.

The third:—

[Compared with a mother's love] the great lake is shallow indeed!
[By this saying] the goodness of a mother should be estimated.

The fourth:—

Even though in poverty we have to pass our days,
Let us never turn aside from the one straight path.

The fifth:—

The person whose heart never changes with time,
A true man or woman that person must be deemed.

---

[1] Lit., "A person having filial piety is called a person." The word *hito* (person), usually indicating either a man or a woman, is often used in the signification of "people" or "Mankind." The full meaning of the sentence is that no unfilial person deserves to be called a human being.

The sixth:—

If the time [of the present] be spent in vain,
In the time of the future must sorrow be borne.

The seventh:—

That a kindness done is not for the sake of others only,
But also for one's own sake, should well be kept in mind.

The eighth:—

Even the sorrow of numberless misfortunes
We shall easily escape if the heart be pure.

The ninth:—

If the heart and the speech be kept straight and true,
The Gods and the Buddhas will surely guard us well.

The tenth:—

In order to become a person held in honor,
As a filial person one must [first] be known.

The reader may think to himself, "How terribly exigent the training that could require the repetition of moral lessons even in a 'ball-play song'!" True,—but it produced perhaps the very sweetest type of woman that this world has ever known.

In some dance-songs the burthen is made by the mere repetition of the last line, or of part of the last line, of each stanza. The following queer ballad exemplifies the practice, and is furthermore remarkable by reason of the curious onomatopoetic choruses introduced at certain passages of the recitative: —

KANÉ-MAKI-ODORI UTA

("Bell-wrapping-dance song." — Province of Iga — Naga district)

A Yamabushi of Kyōto went to Kumano. There resting in the inn Chōjaya, by the beach of Shirotaka, he saw a little girl three years old; and he petted and hugged her, playfully promising to make her his wife, —

(Chorus) *Playfully promising.*

Thereafter that Yamabushi travelled in various provinces; returning only when that girl was thirteen years old. "O my princess, my princess!" he cried to her, — "my little princess, pledged to me by promise!" — "O Sir Yamabushi," made she answer, — "good Sir Yamabushi, take me with you now! —

"Take me with you now!"

"O soon," he said, "I shall come again; soon I shall come again: then, when I come again, I shall take you with me, —

"Take you with me."

Therewith the Yamabushi, escaping from her, quickly, quickly fled away; — with all haste he fled away. Having passed through Tanabé and passed through Minabé, he fled on over the Komatsu moor, —

Over the Komatsu moor.

KAKKARA, KAKKARA, KAKKARA, KAKKA![1]

Therewith the damsel, pursuing, quickly, quickly followed after him; — with all speed she followed after him. Having passed through Tanabé and passed through Minabé, she pursued him over the Komatsu moor, —

Over the Komatsu moor.

Then the Yamabushi, fleeing, came as he fled to the river of Amoda, and cried to the boatman of the river of Amoda, — "O good boatman, good sir boatman, behind me comes a maid pursuing! — pray do not take her across, good boatman, —

"Good sir boatman!"

DEBOKU, DEBOKU, DEBOKU, DENDEN![2]

Then the damsel, pursuing, came to the river of Amoda and called to the boatman, "Bring hither the boat! — take me over in the boat!" — "No, I will not bring the boat; I will not take you over: my boat is forbidden to carry women! —

"Forbidden to carry women!"

"If you do not take me over, I will cross! — if you do not take me over, I will cross! — there is a way to cross the river of Amoda!" Taking off her sandals and holding them aloft, she entered the water, and at once turned into a dragon with twelve horns fully grown, —

With twelve horns fully grown.

---

[1] These syllables, forming a sort of special chorus, are simply onomatopes; intended to represent the sound of sandalled feet running at utmost speed.

[2] These onomatopes, chanted by all the dancers together in chorus, with appropriate gesture, represent the sound of the ferryman's single oar, or scull, working upon its wooden peg. The syllables have no meaning in themselves.

Then the Yamabushi, fleeing, reached the temple Dōjōji, and cried to the priests of the temple Dōjōji: — "O good priests, behind me a damsel comes pursuing! — hide me, I beseech you, good sir priests! —

"Good sir priests!"

Then the priests, after holding consultation, took down from its place the big bell of the temple; and under it they hid him, —

Under it they hid him.

Then the dragon-maid, pursuing, followed him to the temple Dōjōji. For a moment she stood in the gate of the temple: she saw that bell, and viewed it with suspicion. She thought: — "I must wrap myself about it once." She thought: — "I must wrap myself about it twice!" At the third wrapping, the bell was melted, and began to flow like boiling water, —

Like boiling water.

So is told the story of the Wrapping of the Bell. Many damsels dwell by the seashore of Japan; — but who among them, like the daughter of the Chōja, will become a dragon? —

Become a dragon?

This is all the Song of the Wrapping of the Bell! — this is all the Song, —

All the song! [1]

---

[1] This legend forms the subject of several Japanese dramas, both ancient and modern. The original story is that a Buddhist priest, called Anchin, having rashly excited the affection of a maiden named Kiyohimé, and being, by reason of his vows, unable to wed her, sought safety from her advances in flight. Kiyohimé, by the violence of her frustrated passion, therewith became transformed into a fiery dragon; and in that shape she pursued the priest to the temple called Dōjōji, in

I shall give only one specimen of the true street-ballad, — the kind of ballad commonly sung by wandering samisen-players. It is written in an irregular measure, varying from twelve to sixteen syllables in length; the greater number of lines having thirteen syllables. I do not know the date of its composition; but I am told by aged persons who remember hearing it sung when they were children, that it was popular in the period of Tenpō (1830–1843). It is not divided into stanzas; but there are pauses at irregular intervals, — marked by the refrain, *Yanrei!*

O-KICHI-SEIZA KUDOKI

(*"The Ditty of O-Kichi and Seiza"*)

Now hear the pitiful story of two that died for love. —
In Kyōto was the thread-shop of Yoëmon, a merchant

Kumano (modern Kishū), where he tried to hide himself under the great temple-bell. But the dragon coiled herself round the bell, which at once became red-hot, so that the body of the priest was totally consumed.

In this rude ballad Kiyohimé figures only as the daughter of an innkeeper, — the *Chōja*, or rich man of his village; while the priest Anchin is changed into a Yamabushi. The Yamabushi are, or at least were, wandering priests of the strange sect called Shugendo, — itinerant exorcists and diviners, professing both Shinto and Buddhism. Of late years their practices have been prohibited by law; and a real Yamabushi is now seldom to be met with.

The temple Dōjōji is still a famous place of pilgrimage. It is situated not far from Gobō, on the western coast of Kishū. The incident of Anchin and the dragon is said to have occurred in the early part of the tenth century.

known far and near,—a man of much wealth. His business prospered; his life was fortunate. One daughter he had, an only child, by name O-Kichi: at sixteen years she was lovely as a flower. Also he had a clerk in his house, by name Seiza, just in the prime of youth, aged twenty-and-two.

Yanrei!

Now the young man Seiza was handsome; and O-Kichi fell in love with him at sight. And the two were so often together that their secret affection became known; and the matter came to the ears of the parents of O-Kichi; and the parents, hearing of it, felt that such a thing could not be suffered to continue.

Yanrei!

So at last, the mother, having called O-Kichi into a private room, thus spoke to her:—"O my daughter, I hear that you have formed a secret relation with the young man Seiza, of our shop. Are you willing to end that relation at once, and not to think any more about that man, O-Kichi?—answer me, O my daughter."

Yanrei!

"O my dear mother," answered O-Kichi, "what is this that you ask me to do? The closeness of the relation between Seiza and me is the closeness of the relation of the ink to the paper that it penetrates.[1] Therefore, whatever may happen, O mother of mine, to separate from Seiza is more than I can bear."

Yanrei!

[1] Lit.:—"that affinity as-for, ink-and-paper-soaked-like affinity."

Then, the father, having called Seiza to the innermost private room, thus spoke to him:—"I called you here only to tell you this: You have turned the mind of our daughter away from what is right; and even to hear of such a matter is not to be borne. Pack up your things at once, and go!—to-day is the utmost limit of the time that you remain in this house."

Yanrei!

Now Seiza was a native of Ōsaka. Without saying more than "Yes—yes," he obeyed and went away, returning to his home. There he remained four or five days, thinking only of O-Kichi. And because of his longing for her, he fell sick; and as there was no cure and no hope for him, he died.

Yanrei!

Then one night O-Kichi, in a moment of sleep, saw the face of Seiza close to her pillow,—so plainly that she could not tell whether it was real, or only a dream. And rising up, she looked about; but the form of Seiza had vanished.

Yanrei!

Because of this she made up her mind to go at once to the house of Seiza. And, without being seen by any one, she fled from the home of her parents.

Yanrei!

When she came to the ferry at the next village, she did not take the boat, but went round by another road; and making all haste she found her way to the city of Ōsaka. There she asked for the house of Seiza; and she learned that it was in a certain street, the third house from a certain bridge.

Yanrei!

Arriving at last before the home of Seiza, she took off her travelling hat of straw; and seating herself on the threshold of the entrance, she cried out:—"Pardon me kindly!—is not this the house of Master Seiza?"

Yanrei!

Then—O the pity of it!—she saw the mother of Seiza, weeping bitterly, and holding in her hand a Buddhist rosary. "O my good young lady," the mother of Seiza asked, "whence have you come; and whom do you want to see?"

Yanrei!

And O-Kichi said:—"I am the daughter of the thread-merchant of Kyōto. And I have come all the way here only because of the relation that has long existed between Master Seiza and myself. Therefore, I pray you, kindly permit me to see him."

Yanrei!

"Alas!" made answer the mother, weeping, "Seiza, whom you have come so far to see, is dead. To-day is the seventh day from the day on which he died." . . . Hearing these words, O-Kichi herself could only shed tears.

Yanrei!

But after a little while she took her way to the cemetery. And there she found the sotoba[1] erected above the grave of Seiza; and leaning upon it, she wept aloud.

Yanrei!

---

[1] A wooden lath, bearing Buddhist texts, planted above graves. For a full account of the sotoba see my *Exotics and Retrospectives*: "The Literature of the Dead."

Then — how fearful a thing is the longing of a person[1] — the grave of Seiza split asunder; and the form of Seiza rose up therefrom and spoke.

Yanrei!

"Ah! is not this O-Kichi that has come? Kind indeed it was to have come to me from so far away! My O-Kichi, do not weep thus. Never again — even though you weep — can we be united in this world. But as you love me truly, I pray you to set some fragrant flowers before my tomb, and to have a Buddhist service said for me upon the anniversary of my death."

Yanrei!

And with these words the form of Seiza vanished. "O wait, wait for me!" cried O-Kichi, — "wait one little moment![2] I cannot let you return alone! — I shall go with you in a little time!"

Yanrei!

---

[1] In the original: — *Hito no omoi wa osoroshi mono yo!* — ("how fearful a thing is the thinking of a person!"). The word *omoi*, used here in the sense of "longing," refers to the weird power of Seiza's dying wish to see his sweetheart. Even after his burial, this longing has the strength to burst open the tomb.

— In the old English ballad of "William and Marjorie" (see Child: vol. ii. p. 151) there is also a remarkable fancy about the opening and closing of a grave: —

She followed him high, she followed him low,
Till she came to yon churchyard green;
And there the deep grave opened up,
And young William he lay down.

[2] With this episode compare the close of the English ballad "Sweet William's Ghost" (Child: vol. ii., page 148): —

"O stay, my only true love, stay!"
The constant Margaret cried:
Wan grew her cheeks; she closed her een,
Stretched her soft limbs, and died.

Then quickly she went beyond the temple-gate to a moat some four or five *chō*[1] distant; and having filled her sleeves with small stones, into the deep water she cast her forlorn body.

Yanrei!

And now I shall terminate this brief excursion into unfamiliar song-fields by the citation of two Buddhist pieces. The first is from the famous work *Gempei Seisuiki* ("Account of the Prosperity and Decline of the Houses of Gen and Hei"), probably composed during the latter part of the twelfth, or at the beginning of the thirteenth century. It is written in the measure called *Imayō*, — that is to say, in short lines alternately of seven and of five syllables (7, 5; 7, 5; 7, 5, *ad libitum*). The other philosophical composition is from a collection of songs called *Ryūtachi-bushi* ("Ryūtachi Airs"), belonging to the sixteenth century: —

I

(*Measure, Imayō*)

Sama mo kokoro mo
Kawaru kana!
Otsuru namida wa

[1] A *chō* is about one fifteenth of a mile.

Taki no mizu:
Myō-hō-rengé no
Iké to nari;
Guzé no funé ni
Sao sashité;
Shizumu waga mi wo
Nosé-tamaë!

Both form and mind—
Lo! how these change!
The falling of tears
Is like the water of a cataract.
Let them become the Pool
Of the Lotos of the Good Law!
Poling thereupon
The Boat of Salvation,
Vouchsafe that my sinking
Body may ride!

II

(*Period of Bunrokū*—1592–1596)

Who twice shall live his youth?
What flower faded blooms again?
Fugitive as dew
Is the form regretted,
Seen only
In a moment of dream.

FANTASIES

. . . Vainly does each, as he glides,
Fable and dream
Of the lands which the River of Time
Had left ere he woke on its breast,
Or shall reach when his eyes have
    been closed.

MATTHEW ARNOLD

Noctilucæ

Noctilucæ

THE moon had not yet risen; but the vast of the night was all seething with stars, and bridged by a Milky Way of extraordinary brightness. There was no wind; but the sea, far as sight could reach, was running in ripples of fire, — a vision of infernal beauty. Only the ripplings were radiant (between them was blackness absolute); — and the luminosity was amazing. Most of the undulations were yellow like candle-flame; but there were crimson lampings also, — and azure, and orange, and emerald. And the sinuous flickering of all seemed, not a pulsing of many waters, but a laboring of many wills, — a fleeting conscious and monstrous, — a writhing and a swarming incalculable, as of dragon-life in some depth of Erebus.

And life indeed was making the sinister splendor of that spectacle — but life infinitesimal, and of ghostliest delicacy,—life illimitable, yet ephemeral, flaming and fading in ceaseless alternation over the whole round of waters even to the sky-line, above which, in the vaster abyss, other countless lights were throbbing with other spectral colors.

Watching, I wondered and I dreamed. I thought of the Ultimate Ghost revealed in that scintillation tremendous of Night and Sea;—quickening above me, in systems aglow with awful fusion of the past dissolved, with vapor of the life again to be;—quickening also beneath me, in meteor-gushings and constellations and nebulosities of colder fire,—till I found myself doubting whether the million ages of the sun-star could really signify, in the flux of perpetual dissolution, anything more than the momentary sparkle of one expiring noctiluca.

Even with the doubt, the vision changed. I saw no longer the sea of the ancient East, with its shudderings of fire, but that Flood whose width and depth and altitude are one with the Night of Eternity,—the shoreless and timeless Sea of Death and Birth. And the luminous haze of a hundred millions of suns,—the Arch of the Milky Way,—was a single smouldering surge in the flow of the Infinite Tides.

Yet again there came a change. I saw no more that vapory surge of suns; but the living darkness streamed and thrilled about me with infinite sparkling; and every sparkle was beating like a heart,—beating out colors like the tints of the sea-fires. And the lampings of all continually flowed away, as shivering threads of radiance, into illimitable Mystery. . . .

Then I knew myself also a phosphor-point,—one fugitive floating sparkle of the measureless current;—and I saw that the light which was mine shifted tint with each changing of thought. Ruby it sometimes shone, and sometimes sapphire: now it was flame of topaz; again, it was fire of emerald. And the meaning of the changes I could not fully know. But thoughts of the earthly life seemed to make the light burn red; while thoughts of supernal being,—of ghostly beauty and of ghostly bliss,—seemed to kindle ineffable rhythms of azure and of violet.

But of white lights there were none in all the Visible. And I marvelled.

Then a Voice said to me:—

"The White are of the Altitudes. By the blending of the billions they are made. Thy part is to help to their kindling. Even as the color of thy burning, so is the worth of thee. For a moment only is thy quickening; yet the light of thy pulsing lives on: by thy thought, in that shining moment, thou becomest a Maker of Gods."

A Mystery of Crowds

A Mystery of Crowds

WHO has not at some time leaned over the parapet of a bridge to watch the wrinklings and dimplings of the current below, — to wonder at the trembling permanency of surface-shapes that never change, though the substance of them is never for two successive moments the same? The mystery of the spectacle fascinates; and it is worth thinking about. Symbols of the riddle of our own being are those shuddering forms. In ourselves likewise the substance perpetually changes with the flow of the Infinite Stream; but the shapes, though ever agitated by various inter-opposing forces, remain throughout the years.

And who has not been fascinated also by the sight of the human stream that pours and pulses through the streets of some great metropolis? This, too, has its currents and counter-currents and eddyings,—all strengthening or weakening according to the tide-rise or tide-ebb of the city's sea of toil. But the attraction of the greater spectacle for us is not really the mystery of motion: it is rather the mystery of man. As outside observers we are interested chiefly by the passing forms and faces,—by their intimations of personality, their suggestions of sympathy or repulsion. We soon cease to think about the general flow. For the atoms of the human current are visible to our gaze: we see them walk, and deem their movements sufficiently explained by our own experience of walking. And, nevertheless, the motions of the visible individual are more mysterious than those of the always invisible molecule of water.—I am not forgetting the truth that all forms of motion are ultimately incomprehensible: I am referring only to the fact that our common relative knowledge of motions, which are supposed to depend upon will, is even less than our possible relative knowledge of the behavior of the atoms of a water-current.

Every one who has lived in a great city is aware of certain laws of movement which regulate the flow of population through the more crowded thoroughfares. (We need not for present purposes concern ourselves about the complex middle-currents of the living river, with their thunder of hoofs and wheels: I shall speak of the side-currents only.) On either footpath the crowd naturally divides itself into an upward and a downward stream. All persons going in one direction take the right-hand side; all going in the other direction take the left-hand side. By moving with either one of these two streams you can proceed even quickly; but you cannot walk against it: only a drunken or insane person is likely to attempt such a thing. Between the two currents there is going on, by reason of the pressure, a continual self-displacement of individuals to left and right, alternately,—such a yielding and swerving as might be represented, in a drawing of the double-current, by zigzag medial lines ascending and descending. This constant yielding alone makes progress possible: without it the contrary streams would quickly bring each other to a standstill by lateral pressure. But it is especially where two crowd-streams intersect each other, as at street-angles, that this systematic self-displacement is worthy of study. Everybody observes the phenomenon; but few persons think about it. Whoever really thinks about it will discover that there is a mystery in it,—a mystery which no individual experience can fully explain.

In any thronged street of a great metropolis thousands of people are constantly turning aside to left or right in order to pass each other. Whenever two persons walking in contrary directions come face to face in such a press, one of three things is likely to happen:—Either there is a mutual yielding,—or one makes room for the other,—or else both, in their endeavor to be accommodating, step at once in the same direction, and as quickly repeat the blunder by trying to correct it, and so keep dancing to and fro in each other's way,—until the first to perceive the absurdity of the situation stands still, or until the more irritable actually pushes his *vis-à-vis* to one side. But these blunders are relatively infrequent: all necessary yielding, as a rule, is done quickly and correctly.

Of course there must be some general law regulating all this self-displacement,—some law in accord with the universal law of motion in the direction of least resistance. You have only to watch any crowded street for half an hour to be convinced of this. But the law is not easily found or formulated: there are puzzles in the phenomenon.

If you study the crowd-movement closely, you will perceive that those encounters in which one person yields to make way for the other are much less common than those in which both parties give way. But a little reflection will convince you that, even in cases of mutual yielding, one person must of necessity yield sooner than the other,—though the difference in time of the impulse-manifestation should be—as it often is—altogether inappreciable. For the sum of character, physical and psychical, cannot be precisely the same in two human beings. No two persons can have exactly equal faculties of perception and will, nor exactly similar qualities of that experience which expresses itself in mental and physical activities. And therefore in every case of apparent mutual yielding, the yielding must really be successive, not simultaneous. Now although what we might here call the "personal equation" proves that in every case of mutual yielding one individual necessarily yields sooner than the other, it does not at all explain the mystery of the individual impulse in cases where the yielding is not mutual;—it does not explain why you feel at one time that you are about to make your *vis-à-vis* give place, and feel at another time that you must yourself give place. What originates the feeling?

A friend once attempted to answer this question by the ingenious theory of a sort of eye-duel between every two persons coming face to face in a street-throng; but I feel sure that his theory could account for the psychological facts in scarcely half-a-dozen of a thousand such encounters. The greater number of people hurrying by each other in a dense press rarely observe faces: only the disinterested idler has time for that. Hundreds actually pass along the street with their eyes fixed upon the pavement. Certainly it is not the man in a hurry who can guide himself by ocular snap-shot views of physiognomy;—he is usually absorbed in his own thoughts. . . . I have studied my own case repeatedly. While in a crowd I seldom look at faces; but without any conscious observation I am always able to tell when I should give way, or when my *vis-à-vis* is going to save me that trouble. My knowledge is certainly intuitive—a mere knowledge of feeling; and I know not with what to compare it except that blind faculty by which, in absolute darkness, one becomes aware of the proximity of bulky objects without touching them. And my intuition is almost infallible. If I hesitate to obey it, a collision is the invariable consequence.

Furthermore, I find that whenever automatic, or at least semi-conscious, action is replaced by reasoned action—in plainer words, whenever I begin to think about my movements—I always blunder. It is only while I am thinking of other matters,—only while I am acting almost automatically,—that I can thread a dense crowd with ease. Indeed, my personal experience has convinced me that what pilots one quickly and safely through a thick press is not conscious observation at all, but unreasoning, intuitive perception. Now intuitive action of any kind represents inherited knowledge, the experience of past lives,—in this case the experience of past lives incalculable.

Utterly incalculable. . . . Why do I think so? Well, simply because this faculty of intuitive self-direction in a crowd is shared by man with very inferior forms of animal being,—evolutional proof that it must be a faculty immensely older than man. Does not a herd of cattle, a herd of deer, a flock of sheep, offer us the same phenomenon of mutual yielding? Or a flock of birds—gregarious birds especially: crows, sparrows, wild pigeons? Or a shoal of fish? Even among insects—bees, ants, termites—we can study the same law of intuitive self-displacement. The yielding, in all these cases, must still represent an inherited experience unimaginably old. Could we endeavor to retrace the whole course of such inheritance, the attempt would probably lead us back, not only to the very beginnings of sentient life upon this planet, but further,—back into the history of non-sentient substance,—back even to the primal evolution of those mysterious tendencies which are stored up in the atoms of elements. Such atoms we know of only as points of multiple resistance,—incomprehensible knittings of incomprehensible forces. Even the tendencies of atoms doubtless represent accumulations of inheritance——but here thought checks with a shock at the eternal barrier of the Infinite Riddle.

Gothic Horror

Gothic Horror

I

LONG before I had arrived at what catechisms call the age of reason, I was frequently taken, much against my will, to church. The church was very old; and I can see the interior of it at this moment just as plainly as I saw it forty years ago, when it appeared to me like an evil dream. There I first learned to know the peculiar horror that certain forms of Gothic architecture can inspire. . . . I am using the word "horror" in a classic sense, — in its antique meaning of ghostly fear.

On the very first day of this experience, my child-fancy could place the source of the horror. The wizened and pointed shapes of the windows immediately terrified me. In their outline I found the form of apparitions that tormented me in sleep; — and at once I began to imagine some dreadful affinity between goblins and Gothic churches. Presently, in the tall doorways, in the archings of the aisles, in the ribbings and groinings of the roof, I discovered other and wilder suggestions of fear. Even the façade of the organ, — peaking high into the shadow above its gallery, — seemed to me a frightful thing. . . . Had I been then suddenly obliged to answer the question, "What are you afraid of?" I should have whispered, "*Those points!*" I could not have otherwise explained the matter: I only knew that I was afraid of the "points."

Of course the real enigma of what I felt in that church could not present itself to my mind while I continued to believe in goblins. But long after the age of superstitious terrors, other Gothic experiences severally revived the childish emotion in so startling a way as to convince me that childish fancy could not account for the feeling. Then my curiosity was aroused; and I tried to discover some rational cause for the horror. I read many books, and asked many questions; but the mystery seemed only to deepen.

Books about architecture were very disappointing. I was much less impressed by what I could find in them than by references in pure fiction to the awfulness of Gothic art, — particularly by one writer's confession that the interior of a Gothic church, seen at night, gave him the idea of being inside the skeleton of some monstrous animal; and by a far-famed comparison of the windows of a cathedral to eyes, and of its door to a great mouth, "devouring the people." These imaginations explained little; they could not be developed beyond the phase of vague intimation: yet they stirred such emotional response that I felt sure they had touched some truth. Certainly the architecture of a Gothic cathedral offers strange resemblances to the architecture of bone; and the general impression that it makes upon the mind is an impression of life. But this impression or sense of life I found to be indefinable, — not a sense of any life organic, but of a life latent and dæmonic. And the manifestation of that life I felt to be in the *pointing* of the structure.

Attempts to interpret the emotion by effects of altitude and gloom and vastness appeared to me of no worth; for buildings loftier and larger and darker than any Gothic cathedral, but of a different order of architecture, — Egyptian, for instance,—could not produce a like impression. I felt certain that the horror was made by something altogether peculiar to Gothic construction, and that this something haunted the tops of the arches.

"Yes, Gothic architecture is awful," said a religious friend, "because it is the visible expression of Christian faith. No other religious architecture symbolizes spiritual longing; but the Gothic embodies it. Every part climbs or leaps; every supreme detail soars and points like fire. . . ." "There may be considerable truth in what you say," I replied;—"but it does not relate to the riddle that baffles me. Why should shapes that symbolize spiritual longing create horror? Why should any expression of Christian ecstasy inspire alarm? . . ."

Other hypotheses in multitude I tested without avail; and I returned to the simple and savage conviction that the secret of the horror somehow belonged to the points of the archings. But for years I could not find it. At last, at last, in the early hours of a certain tropical morning, it revealed itself quite unexpectedly, while I was looking at a glorious group of palms.

Then I wondered at my stupidity in not having guessed the riddle before.

II

THE characteristics of many kinds of palm have been made familiar by pictures and photographs. But the giant palms of the American tropics cannot be adequately represented by the modern methods of pictorial illustration: they must be seen. You cannot draw or photograph a palm two hundred feet high.

The first sight of a group of such forms, in their natural environment of tropical forest, is a magnificent surprise, — a surprise that strikes you dumb. Nothing seen in temperate zones, — not even the huger growths of the Californian slope, — could have prepared your imagination for the weird solemnity of that mighty colonnade. Each stone-grey trunk is a perfect pillar, — but a pillar of which the stupendous grace has no counterpart in the works of man. You must strain your head well back to follow the soaring of the prodigious column, up, up, up through abysses of green twilight, till at last — far beyond a break in that infinite interweaving of limbs and lianas which is the roof of the forest — you catch one dizzy glimpse of the capital: a parasol of emerald feathers outspread in a sky so blinding as to suggest the notion of azure electricity.

Now what is the emotion that such a vision excites, — an emotion too powerful to be called wonder, too weird to be called delight? Only when the first shock of it has passed, — when the several elements that were combined in it have begun to set in motion widely different groups of ideas, — can you comprehend how very complex it must have been. Many impressions belonging to personal experience were doubtless revived in it, but also with them a multitude of sensations more shadowy, — accumulations of organic memory; possibly even vague feelings older than man, — for the tropical shapes that aroused the emotion have a history more ancient than our race.

One of the first elements of the emotion to become clearly distinguishable is the æsthetic; and this, in its general mass, might be termed the sense of terrible beauty. Certainly the spectacle of that unfamiliar life, — silent, tremendous, springing to the sun in colossal aspiration, striving for light against Titans, and heedless of man in the gloom beneath as of a groping beetle,— thrills like the rhythm of some single marvellous verse that is learned in a glance and remembered forever. Yet the delight, even at its vividest, is shadowed by a queer disquiet. The aspect of that monstrous, pale, naked, smooth-stretching column suggests a life as conscious as the serpent's. You stare at the towering lines of the shape,— vaguely fearing to discern some sign of stealthy movement, some beginning of undulation. Then sight and reason combine to correct the suspicion. Yes, motion is there, and life enormous—but a life seeking only sun,—life, rushing like the jet of a geyser, straight to the giant day.

III

DURING my own experience I could perceive that certain feelings commingled in the wave of delight,—feelings related to ideas of power and splendor and triumph,—were accompanied by a faint sense of religious awe. Perhaps our modern æsthetic sentiments are so interwoven with various inherited elements of religious emotionalism that the recognition of beauty cannot arise independently of reverential feeling. Be this as it may, such a feeling defined itself while I gazed; — and at once the great grey trunks were changed to the pillars of a mighty aisle; and from altitudes of dream there suddenly descended upon me the old dark thrill of Gothic horror.

Even before it died away, I recognized that it must have been due to some old cathedral-memory revived by the vision of those giant trunks uprising into gloom. But neither the height nor the gloom could account for anything beyond the memory. Columns tall as those palms, but supporting a classic entablature, could evoke no sense of disquiet resembling the Gothic horror. I felt sure of this, — because I was able, without any difficulty, to shape immediately the imagination of such a façade. But presently the mental picture distorted. I saw the architrave elbow upward in each of the spaces between the pillars, and curve and point itself into a range of prodigious arches; — and again the sombre thrill descended upon me. Simultaneously there flashed to me the solution of the mystery. I understood that the Gothic horror was *a horror of monstrous motion*, — and that it had seemed to belong to the points of the arches because the idea of such motion was chiefly suggested by the extraordinary angle at which the curves of the arching touched.

To any experienced eye, the curves of Gothic arching offer a striking resemblance to certain curves of vegetal growth;—the curves of the palm-branch being, perhaps, especially suggested. But observe that the architectural form suggests more than any vegetal comparison could illustrate! The meeting of two palm-crests would indeed form a kind of Gothic arch; yet the effect of so short an arch would be insignificant. For nature to repeat the strange impression of the real Gothic arch, it were necessary that the branches of the touching crests should vastly exceed, both in length of curve and strength of spring, anything of their kind existing in the vegetable world. The effect of the Gothic arch depends altogether upon the intimation of energy. An arch formed by the intersection of two short sprouting lines could suggest only a feeble power of growth; but the lines of the tall mediæval arch seem to express a crescent force immensely surpassing that of nature. And the horror of Gothic architecture is not in the mere suggestion of a growing life, but in the suggestion of an energy supernatural and tremendous.

Of course the child, oppressed by the strangeness of Gothic forms, is yet incapable of analyzing the impression received: he is frightened without comprehending. He cannot divine that the points and the curves are terrible to him because they represent the prodigious exaggeration of a real law of vegetal growth. He dreads the shapes because they seem alive; yet he does not know how to express this dread. Without suspecting why, he feels that this silent manifestation of power, everywhere pointing and piercing upward, is not natural. To his startled imagination, the building stretches itself like a phantasm of sleep,—makes itself tall and taller with intent to frighten. Even though built by hands of men, it has ceased to be a mass of dead stone: it is infused with Something that thinks and threatens;—it has become a shadowing malevolence, a multiple goblinry, a monstrous fetish!

Levitation

Levitation

OUT of some upper-story window I was looking into a street of yellow-tinted houses,—a colonial street, old-fashioned, narrow, with palm-heads showing above its roofs of tile. There were no shadows; there was no sun,—only a grey soft light, as of early gloaming.

Suddenly I found myself falling from the window; and my heart gave one sickening leap of terror. But the distance from window to pavement proved to be much greater than I supposed,—so great that, in spite of my fear, I began to wonder. Still I kept falling, falling,—and still the dreaded shock did not come. Then the fear ceased, and a queer pleasure took its place;—for I discovered that I was not falling quickly, but only *floating* down. Moreover, I was floating feet foremost—must have turned in descending. At last I touched the stones — but very, very lightly, with only one foot; and instantly at that touch I went up again, — rose to the level of the eaves. People stopped to stare at me. I felt the exultation of power superhuman; — I felt for the moment as a god.

Then softly I began to sink; and the sight of faces, gathering below me, prompted a sudden resolve to fly down the street, over the heads of the gazers. Again like a bubble I rose, and, with the same impulse, I sailed in one grand curve to a distance that astounded me. I felt no wind; — I felt nothing but the joy of motion triumphant. Once more touching pavement, I soared at a bound for a thousand yards. Then, reaching the end of the street, I wheeled and came back by great swoops, — by long slow aerial leaps of surprising altitude. In the street there was dead silence: many people were looking; but nobody spoke. I wondered what they thought of my feat, and what they would say if they knew how easily the thing was done. By the merest chance I had found out how to do it; and the only reason why it seemed a feat was that no one else had ever attempted it. Instinctively I felt that to say anything about the accident, which had led to the discovery, would be imprudent. Then the real meaning of the strange hush in the street began to dawn upon me. I said to myself: —

"This silence is the Silence of Dreams; — I am quite well aware that this is a dream. I remember having dreamed the same dream before. But the discovery of this power is not a dream: *it is a revelation!* . . . Now that I have learned how to fly, I can no more forget it than a swimmer can forget how to swim. To-morrow morning I shall astonish the people, by sailing over the roofs of the town."

Morning came; and I woke with the fixed resolve to fly out of the window. But no sooner had I risen from bed than the knowledge of physical relations returned, like a sensation forgotten, and compelled me to recognize the unwelcome truth that I had not made any discovery at all.

This was neither the first nor the last of such dreams; but it was particularly vivid, and I therefore selected it for narration as a good example of its class. I still fly occasionally, — sometimes over fields and streams, — sometimes through familiar streets; and the dream is invariably accompanied by remembrance of like dreams in the past, as well as by the conviction that I have really found out a secret, really acquired a new faculty. "This time, at all events," I say to myself, "it is impossible that I can be mistaken;—I *know* that I shall be able to fly after I awake. Many times before, in other dreams, I learned the secret only to forget it on awakening; but this time I am absolutely sure that I shall not forget." And the conviction actually stays with me until I rise from bed, when the physical effort at once reminds me of the formidable reality of gravitation.

The oddest part of this experience is the feeling of buoyancy. It is much like the feeling of floating,—of rising or sinking through tepid water, for example;—and there is no sense of real effort. It is a delight; yet it usually leaves something to be desired. I am a low flyer; I can proceed only like a pteromys or a flying-fish—and far less quickly: moreover, I must tread earth occasionally in order to obtain a fresh impulsion. I seldom rise to a height of more than twenty-five or thirty feet;—the greater part of the time I am merely skimming surfaces. Touching the ground only at intervals of several hundred yards is pleasant skimming; but I always feel, in a faint and watery way, the dead pull of the world beneath me.

Now the experience of most dream-flyers I find to be essentially like my own. I have met but one who claims superior powers: he says that he flies over mountains — goes sailing from peak to peak like a kite. All others whom I have questioned acknowledge that they fly low, — in long parabolic curves, — and this only by touching ground from time to time. Most of them also tell me that their flights usually begin with an imagined fall, or desperate leap; and no less than four say that the start is commonly taken from the top of a stairway.

*
* *

For myriads of years humanity has thus been flying by night. How did the fancied motion, having so little in common with any experience of active life, become a universal experience of the life of sleep?

It may be that memory-impressions of certain kinds of aerial motion, — exultant experiences of leaping or swinging, for example, — are in dream-revival so magnified and prolonged as to create the illusion of flight. We know that in actual time the duration of most dreams is very brief. But in the half-life of sleep — (nightmare offering some startling exceptions) — there is scarcely more than a faint smouldering of consciousness by comparison with the quick flash and vivid thrill of active cerebration; — and time, to the dreaming brain, would seem to be magnified, somewhat as it must be relatively magnified to the feeble consciousness of an insect. Supposing that any memory of the sensation of falling, together with the memory of the concomitant fear, should be accidentally revived in sleep, the dream-prolongation of the sensation and the emotion — unchecked by the natural sequence of shock — might suffice to revive other and even pleasurable memories of airy motion. And these, again, might quicken other combinations of interrelated memories able to furnish all the incident and scenery of the long phantasmagoria.

But this hypothesis will not fully explain certain feelings and ideas of a character different from any experience of waking-hours, — the exultation of voluntary motion without exertion, — the pleasure of the utterly impossible, — the ghostly delight of imponderability. Neither can it serve to explain other dream-experiences of levitation which do not begin with the sensation of leaping or falling, and are seldom of a pleasurable kind. For example, it sometimes happens during nightmare that the dreamer, deprived of all power to move or speak, actually feels his body lifted into the air and floated away by the force of the horror within him. Again, there are dreams in which the dreamer has no physical being. I have thus found myself without any body, — a viewless and voiceless phantom, hovering upon a mountain-road in twilight time, and trying to frighten lonely folk by making small moaning noises. The sensation was of moving through the air by mere act of will: there was no touching of surfaces; and I seemed to glide always about a foot above the road.

Could the feeling of dream-flight be partly interpreted by organic memory of conditions of life more ancient than man, — life weighty, and winged, and flying heavily, *a little above the ground?*

Or might we suppose that some all-permeating Over-Soul, dormant in other time, wakens within the brain at rare moments of our sleep-life? The limited human consciousness has been beautifully compared to the visible solar spectrum, above and below which whole zones of colors invisible await the evolution of superior senses; and mystics aver that something of the ultra-violet or infra-red rays of the vaster Mind may be momentarily glimpsed in dreams. Certainly the Cosmic Life in each of us has been all things in all forms of space and time. Perhaps you would like to believe that it may bestir, in slumber, some vague sense-memory of things more ancient than the sun, — memory of vanished planets with fainter powers of gravitation, where the normal modes of voluntary motion would have been like the realization of our flying dreams? . . .

Nightmare-Touch

Nightmare-Touch

I

WHAT *is* the fear of ghosts among those who believe in ghosts?

All fear is the result of experience, — experience of the individual or of the race, — experience either of the present life or of lives forgotten. Even the fear of the unknown can have no other origin. And the fear of ghosts must be a product of past pain.

Probably the fear of ghosts, as well as the belief in them, had its beginning in dreams. It is a peculiar fear. No other fear is so intense; yet none is so vague. Feelings thus voluminous and dim are super-individual mostly, — feelings inherited, — feelings made within us by the experience of the dead.

What experience?

Nowhere do I remember reading a plain statement of the reason why ghosts are feared. Ask any ten intelligent persons of your acquaintance, who remember having once been afraid of ghosts, to tell you exactly why they were afraid, — to define the fancy behind the fear; — and I doubt whether even one will be able to answer the question. The literature of folk-lore — oral and written — throws no clear light upon the subject. We find, indeed, various legends of men torn asunder by phantoms; but such gross imaginings could not explain the peculiar quality of ghostly fear. It is not a fear of bodily violence. It is not even a reasoning fear, — not a fear that can readily explain itself, — which would not be the case if it were founded upon definite ideas of physical danger. Furthermore, although primitive ghosts may have been imagined as capable of tearing and devouring, the common idea of a ghost is certainly that of a being intangible and imponderable.[1]

[1] I may remark here that in many old Japanese legends and ballads, ghosts are represented as having power to *pull off* people's heads. But so far as the origin of the fear of ghosts is concerned, such stories explain nothing, — since the experiences that evolved the fear must have been real, not imaginary, experiences.

Now I venture to state boldly that the common fear of ghosts is *the fear of being touched by ghosts,*—or, in other words, that the imagined Supernatural is dreaded mainly because of its imagined power to touch. Only to *touch*, remember!—not to wound or to kill.

But this dread of the touch would itself be the result of experience,—chiefly, I think, of prenatal experience stored up in the individual by inheritance, like the child's fear of darkness. And who can ever have had the sensation of being touched by ghosts? The answer is simple:—*Everybody who has been seized by phantoms in a dream.*

Elements of primeval fears—fears older than humanity—doubtless enter into the child-terror of darkness. But the more definite fear of ghosts may very possibly be composed with inherited results of dream-pain,—ancestral experience of nightmare. And the intuitive terror of supernatural touch can thus be evolutionally explained.

Let me now try to illustrate my theory by relating some typical experiences.

II

When about five years old I was condemned to sleep by myself in a certain isolated room, thereafter always called the Child's Room. (At that time I was scarcely ever mentioned by name, but only referred to as "the Child.") The room was narrow, but very high, and, in spite of one tall window, very gloomy. It contained a fire-place wherein no fire was ever kindled; and the Child suspected that the chimney was haunted.

A law was made that no light should be left in the Child's Room at night,—simply because the Child was afraid of the dark. His fear of the dark was judged to be a mental disorder requiring severe treatment. But the treatment aggravated the disorder. Previously I had been accustomed to sleep in a well-lighted room, with a nurse to take care of me. I thought that I should die of fright when sentenced to lie alone in the dark, and—what seemed to me then abominably cruel—actually *locked* into my room, the most dismal room of the house. Night after night when I had been warmly tucked into bed, the lamp was removed; the key clicked in the lock; the protecting light and the footsteps of my guardian receded together. Then an agony of fear would come upon me. Something in the black air would seem to gather and grow—(I thought that I could even *hear* it grow)—till I had to scream. Screaming regularly brought punishment; but it also brought back the light, which more than consoled for the punishment. This fact being at last found out, orders were given to pay no further heed to the screams of the Child.

Why was I thus insanely afraid? Partly because the dark had always been peopled for me with shapes of terror. So far back as memory extended, I had suffered from ugly dreams; and when aroused from them I could always *see* the forms dreamed of, lurking in the shadows of the room. They would soon fade out; but for several moments they would appear like tangible realities. And they were always the same figures. . . . Sometimes, without any preface of dreams, I used to see them at twilight-time,—following me about from room to room, or reaching long dim hands after me, from story to story, up through the interspaces of the deep stairways.

I had complained of these haunters only to be told that I must never speak of them, and that they did not exist. I had complained to everybody in the house; and everybody in the house had told me the very same thing. But there was the evidence of my eyes! The denial of that evidence I could explain only in two ways:— Either the shapes were afraid of big people, and showed themselves to me alone, because I was little and weak; or else the entire household had agreed, for some ghastly reason, to say what was not true. This latter theory seemed to me the more probable one, because I had several times perceived the shapes when I was not unattended; —and the consequent appearance of secrecy frightened me scarcely less than the visions did. Why was I forbidden to talk about what I saw, and even heard,—on creaking stairways,—behind wavering curtains?

"Nothing will hurt you,"—this was the merciless answer to all my pleadings not to be left alone at night. But the haunters *did* hurt me. Only—they would wait until after I had fallen asleep, and so into their power,—for they possessed occult means of preventing me from rising or moving or crying out.

Needless to comment upon the policy of locking me up alone with these fears in a black room. Unutterably was I tormented in that room—for years! Therefore I felt relatively happy when sent away at last to a children's boarding-school, where the haunters very seldom ventured to show themselves.

They were not like any people that I had ever known. They were shadowy dark-robed figures, capable of atrocious self-distortion,—capable, for instance, of growing up to the ceiling, and then across it, and then lengthening themselves, head-downwards, along the opposite wall. Only their faces were distinct; and I tried not to look at their faces. I tried also in my dreams—or thought that I tried—to awaken myself from the sight of them by pulling at my eyelids with my fingers; but the eyelids would remain closed, as if sealed. . . . Many years afterwards, the frightful plates in Orfila's *Traité des Exhumés*, beheld for the first time, recalled to me with a sickening start the dream-terrors of childhood. But to understand the Child's experience, you must imagine Orfila's drawings intensely alive, and continually elongating or distorting, as in some monstrous anamorphosis.

Nevertheless the mere sight of those nightmare-faces was not the worst of the experiences in the Child's Room. The dreams always began with a suspicion, or sensation of something heavy in the air,—slowly quenching will,—slowly numbing my power to move. At such times I usually found myself alone in a large unlighted apartment; and, almost simultaneously with the first sensation of fear, the atmosphere of the room would become suffused, half-way to the ceiling, with a sombre-yellowish glow, making objects dimly visible,—though the ceiling itself remained pitch-black. This was not a true appearance of light: rather it seemed as if the black air were changing color from beneath. . . . Certain terrible aspects of sunset, on the eve of storm, offer like effects of sinister color. . . . Forthwith I would try to escape,—(feeling at every step a sensation *as of wading*),—and would sometimes succeed in struggling half-way across the room;—but there I would always find myself brought to a standstill,—paralyzed by some innominable opposition. Happy voices I could hear in the next room;—I could see light through the transom over the door that I had vainly endeavored to reach;—I knew that one loud cry would save me. But not even by the most frantic effort could I raise my voice above a whisper. . . . And all this signified only that the Nameless was coming, — was nearing, — was mounting the stairs. I could hear the step, — booming like the sound of a muffled drum, — and I wondered why nobody else heard it. A long, long time the haunter would take to come, — malevolently pausing after each ghastly foot-fall. Then, without a creak, the bolted door would open, — slowly, slowly, — and the thing would enter, gibbering soundlessly, — and put out hands, — and clutch me, — and toss me to the black ceiling, — and catch me descending to toss me up again, and again, and again. . . . In those moments the feeling was not fear: fear itself had been torpified by the first seizure. It was a sensation that has no name in the language of the living. For every touch brought a shock of something infinitely worse than pain, — something that thrilled into the innermost secret being of me, — a sort of abominable electricity, discovering unimagined capacities of suffering in totally unfamiliar regions of sentiency. . . . This was commonly the work of a single tormentor; but I can also remember having been caught by a group, and tossed from one to another,—seemingly for a time of many minutes.

III

WHENCE the fancy of those shapes? I do not know. Possibly from some impression of fear in earliest infancy; possibly from some experience of fear in other lives than mine. That mystery is forever insoluble. But the mystery of the shock of the touch admits of a definite hypothesis.

First, allow me to observe that the experience of the sensation itself cannot be dismissed as "mere imagination." Imagination means cerebral activity: its pains and its pleasures are alike inseparable from nervous operation, and their physical importance is sufficiently proved by their physiological effects. Dream-fear may kill as well as other fear; and no emotion thus powerful can be reasonably deemed undeserving of study.

One remarkable fact in the problem to be considered is that the sensation of seizure in dreams differs totally from all sensations familiar to ordinary waking life. Why this differentiation? How interpret the extraordinary massiveness and depth of the thrill?

I have already suggested that the dreamer's fear is most probably not a reflection of relative experience, but represents the incalculable total of ancestral experience of dream-fear. If the sum of the experience of active life be transmitted by inheritance, so must likewise be transmitted the summed experience of the life of sleep. And in normal heredity either class of transmissions would probably remain distinct.

Now, granting this hypothesis, the sensation of dream-seizure would have had its beginnings in the earliest phases of dream-consciousness,—long prior to the apparition of man. The first creatures capable of thought and fear must often have dreamed of being caught by their natural enemies. There could not have been much imagining of pain in these primal dreams. But higher nervous development in later forms of being would have been accompanied with larger susceptibility to dream-pain. Still later, with the growth of reasoning-power, ideas of the supernatural would have changed and intensified the character of dream-fear. Furthermore, through all the course of evolution, heredity would have been accumulating the experience of such feeling. Under those forms of imaginative pain evolved through reaction of religious beliefs, there would persist some dim survival of savage primitive fears, and again, under this, a dimmer but incomparably deeper substratum of ancient animal-terrors. In the dreams of the modern child all these latencies might quicken,—one below another,—unfathomably,—with the coming and the growing of nightmare.

It may be doubted whether the phantasms of any particular nightmare have a history older than the brain in which they move. But the shock of the touch would seem to indicate *some point of dream-contact with the total race-experience of shadowy seizure.* It may be that profundities of Self,—abysses never reached by any ray from the life of sun,—are strangely stirred in slumber, and that out of their blackness immediately responds a shuddering of memory, measureless even by millions of years.

Readings from a Dream-book

Readings from a Dream-book

OFTEN, in the blind dead of the night, I find myself reading a book, — a big broad book, — a dream-book. By " dream-book," I do not mean a book about dreams, but a book made of the stuff that dreams are made of.

I do not know the name of the book, nor the name of its author: I have not been able to see the title-page; and there is no running title. As for the back of the volume, it remains, — like the back of the Moon, — invisible forever.

At no time have I touched the book in any way, — not even to turn a leaf. Somebody, always viewless, holds it up and open before me in the dark; and I can read it only because it is lighted by a light that comes from nowhere. Above and beneath and on either side of the book there is darkness absolute; but the pages seem to retain the yellow glow of lamps that once illuminated them.

A queer fact is that I never see the entire text of a page at once, though I see the whole page itself plainly. The text rises, or seems to rise, to the surface of the paper as I gaze, and fades out almost immediately after having been read. By a simple effort of will, I can recall the vanished sentences to the page; but they do not come back in the same form as before: they seem to have been oddly revised during the interval. Never can I coax even one fugitive line to reproduce itself exactly as it read at first. But I can always force something to return; and this something remains sharply distinct during perusal. Then it turns faint grey, and appears to sink — as through thick milk — backward out of sight.

By regularly taking care to write down, immediately upon awakening, whatever I could remember reading in the dream-book, I found myself able last year to reproduce portions of the text. But the order in which I now present these fragments is not at all the order in which I recovered them. If they seem to have any interconnection, this is only because I tried to arrange them in what I imagined to be the rational sequence. Of their original place and relation, I know scarcely anything. And, even regarding the character of the book itself, I have been able to discover only that a great part of it consists of dialogues about the Unthinkable.

Fr. I

. . . Then the Wave prayed to remain a wave forever.

The Sea made answer :—

" Nay, thou must break : there is no rest in me. Billions of billions of times thou wilt rise again to break, and break to rise again."

The Wave complained :—

" I fear. Thou sayest that I shall rise again. But when did ever a wave return from the place of breaking ? "

The Sea responded :—

" Times countless beyond utterance thou hast broken ; and yet thou art ! Behold the myriads of the waves that run before thee, and the myriads that pursue behind thee !—all have been to the place of breaking times unspeakable ; and thither they hasten now to break again. Into me they melt, only to swell anew. But pass they must; for there is not any rest in me."

Murmuring, the Wave replied: —

"Shall I not be scattered presently to mix with the mingling of all these myriads? How should I rise again? Never, never again can I become the same."

"The same thou never art," returned the Sea, "at any two moments in thy running: perpetual change is the law of thy being. What is thine 'I'? Always thou art shaped with the substance of waves forgotten, — waves numberless beyond the sands of the shores of me. In thy multiplicity what art thou? — a phantom, an impermanency!"

"Real is pain," sobbed the Wave, — "and fear and hope, and the joy of the light. Whence and what are these, if I be not real?"

"Thou hast no pain," the Sea responded, — "nor fear nor hope nor joy. Thou art nothing — save in me. I am thy Self, thine 'I': thy form is my dream; thy motion is my will; thy breaking is my pain. Break thou must, because there is no rest in me; but thou wilt break only to rise again, — for death is the Rhythm of Life.

Lo! I, too, die that I may live: these my waters have passed, and will pass again, with wrecks of innumerable worlds to the burning of innumerable suns. I, too, am multiple unspeakably: dead tides of millions of oceans revive in mine ebb and flow. Suffice thee to learn that only because thou wast thou art, and that because thou art thou wilt become again."

Muttered the Wave,—

"I cannot understand."

Answered the Sea,—

"Thy part is to pulse and pass,—never to understand. I also,—even I, the great Sea,—do not understand. . . ."

Fr. II

. . . "The stones and the rocks have felt; the winds have been breath and speech; the rivers and oceans of earth have been locked into chambers of hearts. And the palingenesis cannot cease till every cosmic particle shall have passed through the uttermost possible experience of the highest possible life."

"But what of the planetary core?—has that, too, felt and thought?"

" Even so surely as that all flesh has been sun-fire! In the ceaseless succession of integrations and dissolutions, all things have shifted relation and place numberless billions of times. Hearts of old moons will make the surface of future worlds. . . ."

Fr. III

. . . "No regret is vain. It is sorrow that spins the thread, — softer than moonshine, thinner than fragrance, stronger than death, — the Gleipnir-chain of the Greater Memory. . . .

" In millions of years you will meet again; — and the time will not seem long; for a million years and a moment are the same to the dead. Then you will not be all of your present self, nor she be all that she has been: both of you will at once be less, and yet incomparably more. Then, to the longing that must come upon you, body itself will seem but a barrier through which you would leap to her — or, it may be, to him; for sex will have shifted numberless times ere then. Neither will remember; but each will be filled with a feeling immeasurable of having met before. . . ."

Fr. IV

. . . "So wronging the being who loves,— the being blindly imagined but of yesterday,— this mocker mocks the divine in the past of the Soul of the World. Then in that heart is revived the countless million sorrows buried in forgotten graves,—all the old pain of Love, in its patient contest with Hate, since the beginning of Time.

"And the Gods know,—the dim ones who dwell beyond Space,—spinning the mysteries of Shape and Name. For they sit at the roots of Life; and the pain runs back to them; and they feel that wrong,—as the Spider feels in the trembling of her web that a thread is broken. . . ."

Fr. V

. . . "Love at sight is the choice of the dead. But the most of them are older than ethical systems; and the decision of their majorities is rarely moral. They choose by beauty,—according to their memory of physical excellence; and as bodily fitness makes the foundation of mental and of moral power, they are not apt to choose ill. Nevertheless they are sometimes strangely cheated. They have been known to want beings that could never help ghost to a body, — hollow goblins. . . ."

Fr. VI

. . . "The Animulæ making the Self do not fear death as dissolution. They fear death only as reintegration, — recombination with the strange and the hateful of other lives: they fear the imprisonment, within another body, of that which loves together with that which loathes. . . ."

Fr. VII

. . . "In other time the El-Woman sat only in waste places, and by solitary ways. But now in the shadows of cities she offers her breasts to youth; and he whom she entices, presently goes mad, and becomes, like herself, a hollowness. For the higher ghosts that entered into the making of him perish at that goblin-touch, — die as the pupa dies in the cocoon, leaving only a shell and dust behind. . . ."

Fr. VIII

. . . The Man said to the multitude remaining of his Souls: —

"I am weary of life."

And the remnant replied to him: —

"We also are weary of the shame and pain of dwelling in so vile a habitation. Continually we strive that the beams may break, and the pillars crack, and the roof fall in upon us."

"Surely there is a curse upon me," groaned the Man. "There is no justice in the Gods!"

Then the Souls tumultuously laughed in scorn, — even as the leaves of a wood in the wind do chuckle all together. And they made answer to him: —

"As a fool thou liest! Did any save thyself make thy vile body? Was it shapen — or misshapen — by any deeds or thoughts except thine own?"

"No deed or thought can I remember," returned the Man, "deserving that which has come upon me."

"Remember!" laughed the Souls. "No—the folly was in other lives. But we remember; and remembering, we hate."

"Ye are all one with me!" cried the Man,—"how can ye hate?"

"One with thee," mocked the Souls,—"as the wearer is one with his garment! . . . How can we hate? As the fire that devours the wood from which it is drawn by the fire-maker—even so we can hate."

"It is a cursed world!" cried the Man—"why did ye not guide me?"

The Souls replied to him:—

"Thou wouldst not heed the guiding of ghosts that were wiser than we. . . . Cowards and weaklings curse the world. The strong do not blame the world: it gives them all that they desire. By power they break and take and keep. Life for them is a joy, a triumph, an exultation. But creatures without power merit nothing; and nothingness becomes their portion. Thou and we shall presently enter into nothingness."

"Do ye fear?"—asked the Man.

"There is reason for fear," the Souls answered. "Yet no one of us would wish to delay the time of what we fear by continuing to make part of such an existence as thine."

"But ye have died innumerable times?"—wonderingly said the Man.

"No, we have not," said the Souls,—"not even once that we can remember; and our memory reaches back to the beginnings of this world. We die only with the race."

The Man said nothing,—being afraid. The Souls resumed:—

"Thy race ceases. Its continuance depended upon thy power to serve our purposes. Thou hast lost all power. What art thou but a charnel-house, a mortuary-pit? Freedom we needed, and space: here we have been compacted together, a billion to a pin-point! Doorless our chambers and blind;—and the passages are blocked and broken;—and the stairways lead to nothing. Also there are Haunters here, not of our kind,—Things never to be named."

For a little time the Man thought gratefully of death and dust. But suddenly there came into his memory a vision of his enemy's face, with a wicked smile upon it. And then he wished for longer life,—a hundred years of life and pain,—only to see the grass grow tall above the grave of that enemy. And the Souls mocked his desire:—

"Thine enemy will not waste much thought upon thee. He is no half-man,—thine enemy! The ghosts in that body have room and great light. High are the ceilings of their habitation; wide and clear the passageways; luminous the courts and pure. Like a fortress excellently garrisoned is the brain of thine enemy;—and to any point thereof the defending hosts can be gathered for battle in a moment together. *His* generation will not cease—nay! that face of his will multiply throughout the centuries! Because thine enemy in every time provided for the needs of his higher ghosts: he gave heed to their warnings; he pleasured them in all just ways; he did not fail in reverence to them. Wherefore they now have power to help him at his need. . . . How hast thou reverenced or pleasured us?"

The Man remained silent for a space. Then, as in horror of doubting, he questioned:—

"Wherefore should ye fear—if nothingness be the end?"

"What is nothingness?" the Souls responded. "Only in the language of delusion is there an end. That which thou callest the end is in truth but the very beginning. The essence of us cannot cease. In the burning of worlds it cannot be consumed. It will shudder in the cores of great stars; — it will quiver in the light of other suns. And once more, in some future cosmos, it will reconquer knowledge — but only after evolutions unthinkable for multitude. Even out of the nameless beginnings of form, and thence through every cycle of vanished being, — through all successions of exhausted pain, — through all the Abyss of the Past, — it must climb again."

The Man uttered no word: the Souls spoke on: —

"For millions of millions of ages must we shiver in tempests of fire: then shall we enter anew into some slime primordial, — there to quicken, and again writhe upward through all foul dumb blind shapes. Innumerable the metamorphoses! — immeasurable the agonies! . . . And the fault is not of any Gods: it is thine!"

"Good or evil," muttered the Man, — "what signifies either? The best must become as the worst in the grind of the endless change."

"Nay!" cried out the Souls; "for the strong there is a goal, — the goal that thou couldst not strive to gain. They will help to the fashioning of fairer worlds;—they will win to larger light;—they will tower and soar as flame to enter the Zones of the Divine. But thou and we go back to slime! Think of the billion summers that might have been for us!—think of the joys, the loves, the triumphs cast away!—the dawns of the knowledge undreamed,—the glories of sense unimagined,—the exultations of illimitable power! . . . think, think, O fool, of all that thou hast lost!"

Then the Souls of the Man turned themselves into worms, and devoured him.

In a Pair of Eyes

In a Pair of Eyes

THERE is one adolescent moment never to be forgotten, — the moment when the boy learns that this world contains nothing more wonderful than a certain pair of eyes. At first the surprise of the discovery leaves him breathless: instinctively he turns away his gaze. That vision seemed too delicious to be true. But presently he ventures to look again, — fearing with a new fear, — afraid of the reality, afraid also of being observed; — and lo! his doubt dissolves in a new shock of ecstasy. Those eyes are even more wonderful than he had imagined — nay! they become more and yet more entrancing every successive time that he looks at them! Surely in all the universe there cannot be another such pair of eyes! What can lend them such enchantment? Why do they appear divine? . . . He feels that he must ask somebody to explain, — must propound to older and wiser heads the riddle of his new emotions. Then he makes his confession, with a faint intuitive fear of being laughed at, but with a strange, fresh sense of rapture in the telling. Laughed at he is — tenderly; but this does not embarrass him nearly so much as the fact that he can get no answer to his question, — to the simple "Why?" made so interesting by his frank surprise and his timid blushes. No one is able to enlighten him; but all can sympathize with the bewilderment of his sudden awakening from the long soul-sleep of childhood.

Perhaps that "Why?" never can be fully answered. But the mystery that prompted it constantly tempts one to theorize; and theories may have a worth independent of immediate results. Had it not been for old theories concerning the Unknowable, what should we have been able to learn about the Knowable? Was it not while in pursuit of the Impossible that we stumbled upon the undreamed-of and infinitely marvellous Possible?

Why indeed should a pair of human eyes appear for a time to us so beautiful that, when likening their radiance to splendor of diamond or amethyst or emerald, we feel the comparison a blasphemy? Why should we find them deeper than the sea, deeper than the day, — deep even as the night of Space, with its scintillant mist of suns? Certainly not because of mere wild fancy. These thoughts, these feelings, must spring from some actual perception of the marvellous, — some veritable revelation of the unspeakable. There is, in very truth, one brief hour of life during which the world holds for us nothing so wonderful as a pair of eyes. And then, while looking into them, we discover a thrill of awe vibrating through our delight, — awe made by a something *felt* rather than seen: a latency, — a power, — a shadowing of depth unfathomable as the cosmic Ether. It is as though, through some intense and sudden stimulation of vital being, we had obtained — for one supercelestial moment — the glimpse of a reality, never before imagined, and never again to be revealed.

There is, indeed, an illusion. We seem to view the divine; but this divine itself, whereby we are dazzled and duped, is a ghost. Not to actuality belongs the spell, — not to anything that *is*, — but to some infinite composite phantom of what has been. Wondrous the vision —

but wondrous only because our mortal sight then pierces beyond the surface of the present into profundities of myriads of years, — pierces beyond the mask of life into the enormous night of death. For a moment we are made aware of a beauty and a mystery and a depth unutterable: then the Veil falls again forever.

The splendor of the eyes that we worship belongs to them only as brightness to the morning-star. It is a reflex from beyond the shadow of the Now, — a ghost-light of vanished suns. Unknowingly within that maiden-gaze we meet the gaze of eyes more countless than the hosts of heaven, — eyes otherwhere passed into darkness and dust.

Thus, and only thus, the depth of that gaze is the depth of the Sea of Death and Birth, — and its mystery is the World-Soul's vision, watching us out of the silent vast of the Abyss of Being.

Thus, and only thus, do truth and illusion mingle in the magic of eyes, — the spectral past suffusing with charm ineffable the apparition of the present; — and the sudden splendor in the soul of the Seër is but a flash, — one soundless sheet-lightning of the Infinite Memory.

Other TUT BOOKS available:

BACHELOR'S HAWAII *by Boye de Mente*

BACHELOR'S JAPAN *by Boye de Mente*

BACHELOR'S MEXICO *by Boye de Mente*

A BOOK OF NEW ENGLAND LEGENDS AND FOLK LORE *by Samuel Adams Drake*

THE BUDDHA TREE *by Fumio Niwa; translated by Kenneth Strong*

CALABASHES AND KINGS: An Introduction to Hawaii *by Stanley D. Porteus*

CHINA COLLECTING IN AMERICA *by Alice Morse Earle*

CHINESE COOKING MADE EASY *by Rosy Tseng*

CHOI OI!: The Lighter Side of Vietnam *by Tony Zidek*

CONFUCIUS SAY *by Leo Shaw*

THE COUNTERFEITER and Other Stories *by Yasushi Inoue; translated by Leon Picon*

CURIOUS PUNISHMENTS OF BYGONE DAYS *by Alice Morse Earle*

CUSTOMS AND FASHIONS IN OLD NEW ENGLAND *by Alice Morse Earle*

DINING IN SPAIN *by Gerrie Beene and Lourdes Miranda King*

EXOTICS AND RETROSPECTIVES *by Lafcadio Hearn*

FIRST YOU TAKE A LEEK: A Guide to Elegant Eating Spiced with Culinary Capers *by Maxine J. Saltonstall*

FIVE WOMEN WHO LOVED LOVE *by Saikaku Ihara; translated by William Theodore de Bary*

A FLOWER DOES NOT TALK: Zen Essays *by Abbot Zenkei Shibayama of the Nanzenji*

FOLK LEGENDS OF JAPAN *by Richard M. Dorson*

GLEANINGS IN BUDDHA-FIELDS: Studies of Hand and Soul in the Far East *by Lafcadio Hearn*

GOING NATIVE IN HAWAII: A Poor Man's Guide to Paradise *by Timothy Head*

HAIKU IN ENGLISH *by Harold G. Henderson*

HARP OF BURMA *by Michio Takeyama; translated by Howard Hibbett*

THE HAWAIIAN GUIDE BOOK for Travelers *by Henry M. Whitney*

HAWAIIAN PHRASE BOOK

HAWAII: End of the Rainbow *by Kazuo Miyamoto*

HISTORIC MANSIONS AND HIGHWAYS AROUND BOSTON *by Samuel Adams Drake*

HISTORICAL AND GEOGRAPHICAL DICTIONARY OF JAPAN *by E. Papinot*

A HISTORY OF JAPANESE LITERATURE *by W. G. Aston*

HOMEMADE ICE CREAM AND SHERBERT *by Sheila MacNiven Cameron*

HOW TO READ CHARACTER: A New Illustrated Handbook of Phrenology and Physiognomy, for Students and Examiners *by Samuel R. Wells*

INDIAN RIBALDRY *by Randor Guy*

IN GHOSTLY JAPAN *by Lafcadio Hearn*

JAPAN: An Attempt at Interpretation *by Lafcadio Hearn*

THE JAPANESE ABACUS *by Takashi Kojima*

THE JAPANESE ARE LIKE THAT *by Ichiro Kawasaki*

JAPANESE ETIQUETTE: An Introduction *by the World Fellowship Committee of the Tokyo Y.W.C.A.*

THE JAPANESE FAIRY BOOK *compiled by Yei Theodora Ozaki*

JAPANESE FOLK-PLAYS: The Ink-Smeared Lady and Other Kyogen *translated by Shio Sakanishi*

JAPANESE FOOD AND COOKING *by Stuart Griffin*

JAPANESE HOMES AND THIER SURROUNDINGS *by Edward S. Morse*

A JAPANESE MISCELLANY *by Lafcadio Hearn*

JAPANESE RECIPES *by Tatsuji Tada*

JAPANESE TALES OF MYSTERY & IMAGINATION *by Edogawa Rampo; translated by James B. Harris*

JAPANESE THINGS: Being Notes on Various Subjects Connected with Japan *by Basil Hall Chamberlain*

THE JOKE'S ON JUDO *by Donn Draeger and Ken Tremayne*

THE KABUKI HANDBOOK *by Aubrey S. Halford and Giovanna M. Halford*

KAPPA *by Ryūnosuke Akutagawa; translated by Geoffrey Bownas*

KOKORO: Hints and Echoes of Japanese Inner Life *by Lafcadio Hearn*

KOREAN FOLK TALES *by Im Bang and Yi Ryuk; translated by James S. Gale*

KOTTŌ: Being Japanese Curios, with Sundry Cobwebs *by Lafcadio Hearn*

KWAIDAN: Stories and Studies of Strange Things *by Lafcadio Hearn*

LET'S STUDY JAPANESE *by Jun Maeda*

THE LIFE OF BUDDHA *by A. Ferdinand Herold*

MODERN JAPANESE PRINTS: A Contemporary Selection *edited by Yuji Abe*

MORE ZILCH: The Marine Corps' Most Guarded Secret *by Roy Delgado*

NIHONGI: Chronicles of Japan from the Earliest Times to A.D. 697 *by W. G. Aston*

OLD LANDMARKS AND HISTORIC PERSONAGES OF BOSTON *by Samuel Adams Drake*

ORIENTAL FORTUNE TELLING *by Jimmei Shimano; translated by Togo Taguchi*

PHYSICAL FITNESS: A Practical Program *by Clark Hatch*

POO POO MAKE PRANT GLOW *by Harvey Ward*

PROFILES OF MODERN AMERICAN AUTHORS *by Bernard Dekle*

READ JAPANESE TODAY *by Len Walsh*

SALMAGUNDI VIETNAM *by Don Pratt and Lee Blair*

SELF DEFENSE SIMPLIFIED IN PICTURES *by Don Hepler*

SHADOWINGS *by Lafcadio Hearn*

A SHORT SYNOPSIS OF THE MOST ESSENTIAL POINTS IN HAWAIIAN GRAMMAR *by W. D. Alexander*

THE STORY BAG: A Collection of Korean Folk Tales *by Kim So-un; translated by Setsu Higashi*

SUMI-E: An Introduction to Ink Painting *by Nanae Momiyama*

SUN-DIALS AND ROSES OF YESTERDAY *by Alice Morse Earle*

THE TEN FOOT SQUARE HUT AND TALES OF THE HEIKE: Being Two Thirteenth-century Japanese classics, the "Hojoki" and selections from the "Heike Monogatari" *translated by A. L. Sadler*

THIS SCORCHING EARTH *by Donald Richie*

TIMES-SQUARE SAMURAI or the Improbable Japanese Occupation of New York *by Robert B. Johnson and Billie Niles Chadbourne*

TO LIVE IN JAPAN *by Mary Lee O'Neal and Virginia Woodruff*

THE TOURIST AND THE REAL JAPAN *by Boye de Mente*

TOURS OF OKINAWA: A Souvenir Guide to Places of Interest *compiled by Gasei Higa, Isamu Fuchaku, and Zenkichi Toyama*